Comparative Pathobiology

Volume 4
INVERTEBRATE MODELS
FOR BIOMEDICAL
RESEARCH

Comparative Pathobiology

Comparative Pathobiology

Volume 4
INVERTEBRATE MODELS
FOR BIOMEDICAL
RESEARCH

Edited by **Lee A. Bulla, Jr.**

United States Department of Agriculture
Manhattan, Kansas

and

Thomas C. Cheng

Lehigh University
Bethlehem, Pennsylvania

Springer Science+Business Media, LLC

Library of Congress Cataloging in Publication Data

Main entry under title:

Invertebrate models for biomedical research.

   (Comparative pathobiology; v. 4)
   Papers presented at two symposia held Aug. 21—26, 1977 at the 10th
annual meeting of the Society for Invertebrate Pathology of Michigan State
University, East Lansing.
   Includes index.
   1. Diseases — Animal models — Congresses. 2. Invertebrates — Physiology
— Congresses. 3. Invertebrates — Diseases — Congresses. 4. Animal experi-
mentation — Congresses. I. Bulla, Lee A. II. Cheng, Thomas Clement. III.
Society for Invertebrate Pathology. IV. Series.
RB125.158   619   78-10155

ISBN 978-1-4757-1280-3   ISBN 978-1-4757-1278-0 (eBook)
DOI 10.1007/978-1-4757-1278-0

Acknowledgments

We thank Denise Dalley and Aileen Berroth, whose skillful attention to the typing and assembly of the material in this volume helped make the publication possible.

FOREWORD

On August 21-26, 1977, two symposia were included in the program
of the 10th Annual Meeting of the Society for Invertebrate
Pathology held at Michigan State University, East Lansing,
Michigan. One was entitled "Invertebrate Models for Biomedical
Research" organized by Dr. Thomas C. Cheng, and the second,
organized by Dr. Robert S. Anderson, was entitled "Cellular and
Humoral Reactions to Disease by Invertebrate Animals." When the
final manuscripts of the speakers were received, it became apparent
that all of the papers were so closely related that the editors
decided that they should be combined and published in a single
volume of *Comparative Pathobiology* under the subtitle of
Invertebrate Models for Biomedical Research. This volume is the
result.

We hope that volume four will provide the reader further insight
into the complexity and comprehensiveness of pathobiology.
Pathobiology encompasses not only the study of pathologic conditions
but also the biology of causative agents and response reactions.
Obviously, the diversity of problems related to invertebrate
disease and disease vectors requires fundamental knowledge in
microbiology, parasitology, and invertebrate physiology, biochemistry,
immunology, and development. Such studies cut across the boundaries
of biomedicine, agriculture, and environmental sciences. Although
volume four is directed toward biomedicine, it is our intent that
the collection of past and future volumes of *Comparative Pathobiology*
will provide broad coverage of the variety of research problems in
invertebrate pathobiology and will stimulate new ideas for
researchers who are working with invertebrates and for those investigators who are concerned with the implications of invertebrate
vectors and causative agents in vertebrate diseases.

Lee A. Bulla, Jr.

Thomas C. Cheng

Editors

Preface

The theme of this volume centers on the utilization of invertebrates as experimental models in biomedical research. In the
first contribution, Dr. Burton J. Bogitsh has pointed out that the
esophageal glands of *Schistosoma mansoni* schistosomules are
excellent models for studying the effects of colchicine and
vinblastine on secretion and how this activity is correlated with
feeding and digestion. Furthermore, he has demonstrated that the
inhibition of secretion of the gland by colchicine and vinblastine
is consistent with the theory of microtubule-directed secretory
activities. Therefore, these glands of schistosomes could be
employed with profit for the further elucidation of the mechanisms
underlying cellular secretion.

In the second contribution, Dr. Robert S. Anderson has reported
that several species of invertebrates are capable of metabolizing
benzo[a]pyrene, which is a ubiquitous environmental pollutant
known to be a potent carcinogen for mammals. Therefore,
benzo[a]pyrene may be considered a potential invertebrate carcinogen.
Like in mammals, benzo[a]pyrene is matabolized by the invertebrates
studied via a microsomal mixed-function oxidase system; therefore,
these invertebrates could be useful as models for probing the
carcinogenic mechanism of this compound.

In the third contribution, Dr. George P. Hoskin has presented a
thorough and critical review of what is known about lipid metabolism
in molluscs and has pointed out where lipid metabolism in molluscs
is either similar or different from that in mammals. This, of
course, is useful information for those who wish to employ molluscs
as models for studying lipid metabolism.

The next contribution is by Dr. Thomas C. Cheng. In addition to
a review of what is known about fluctuations in the quantities of
lysosomal enzymes in molluscan phagocytes, he has contributed
evidence that these enzymes play a role in internal defense against
certain bacteria. Also, he has contributed several hypotheses as
to why certain helminth parasites are encapsulated and destroyed
in molluscs while others are not.

In the fifth contribution, Dr. M. R. Tripp and Ms. R. M. Turner
have described the cellular response in the mussel *Mytilus edulis*
to the parasite *Proctoeces maculatus* as well as the seasonal
incidence of this progenetic trematode in its molluscan host.

The sixth contribution is by Drs. T. M. Rizki and Rose M. Rizki.
By employing both scanning and transmission electron microscopy,
they have demonstrated that in the larvae of both the *tu-W* and
tu-ts^82 strains of *Drosophila*, hemocytes respond to degenerative
changes in the fat body cells by forming melanized capsules.
Furthermore, morphological transformation of plasmatocytes preceeds
changes at the prospective tumor-forming site and must therefore
be independent of contact with that site. In contrast, the
encapsulation response of the hemocytes follows surface changes
at the tumor-forming site; therefore, surface contact must play a
role in hemocyte recognition of these degenerating sites.

The seventh chapter, by Dr. Thomas C. Cheng, is concerned with
granuloma formation in incompatible and partially incompatible
strains of the gastropod *Biomphalaria glabrata* infected with
Schistosoma mansoni. By employing acid phosphatase as a marker,
it has been ascertained that each granuloma is comprised of
granulocytes. Furthermore, such foci represent a nonspecific
reaction to foreign bodies which do not have an immunologic basis
sensu strictu.

In the eighth chapter, Dr. David A. Foley describes the types
of hemocytes of the mosquito *Anopheles stephensi*. Also, the
in vitro and *in vivo* interaction between hemocytes with the malaria-
causing protozoan *Plasmodium berghei* as well as with bacteria and
human erythrocytes are described both qualitatively and quantitatively.

In the final contribution, Drs. H. G. Boman, I. Faye, A. Pye, and
T. Rasmuson have described and partially characterized an inducible
humoral factor in giant silk moths. This immunity manifests itself
as a potent, cell-free antibacterial activity in the hemolymph.

This collection of papers adds to our knowledge of the basic
biology of invertebrates of biomedical importance. Invertebrate
pathobiology has been primarily a descriptive subject but, as is
evident from volume four, it is rapidly becoming analytical.
Developments in molecular biology and biochemistry are essential to
an understanding of invertebrate disease processes, and these areas
of biology will continue to be emphasized in future volumes.

Thomas C. Cheng

CONTENTS

Use of Schistosomes for Pharmacological Research[1]

BURTON J. BOGITSH

Department of Biology
Vanderbilt University
Nashville, Tennessee

[1]Supported in part by a grant from the Edna McConnell Clark
Foundation (No. 275-0041)

I. INTRODUCTION

Human schistosomiasis is an illness that infects and debilitates
in excess of 200 million people in more than 70 countries in the
tropical areas of the world. It is small wonder that massive
amounts of money and time are consumed in seeking means to treat,
control, and prevent this widespread scourge upon mankind. Economic conditions in the areas of the world where this condition is
most prevalent are such that there is a constant search for less
costly, more effective means of treatment. One aspect of that
research is the seeking out of new target systems within the
schistosomes which may be vulnerable to chemotherapy and, hence,
provide another means of attack upon this agent. To date, chemotherapeutic attack has focused on three primary targets in the
causative organism: portions of the glycolytic pathway, egg
formation, and the nervous system (see Senft, 1969, and Cheng,
1977 for reviews). However, the exact mechanisms by which most
drugs affect the particular target systems are not well understood,
and are, for the most part, empirical in that effects have been
widely observed but only sketchily explained. Ideally, a chemotherapeutic agent should exert an adverse effect upon some critical
biochemical mechanism (target) in the parasite but have little or
no simultaneous effect upon the human host.

The purpose of this paper is to propose another system within the
schistosomes which might serve as a target for chemotherapeutic
assault, i.e., the microtubule – dependent secretory system. The
structure of the covering of the outer surface, or tegument, of
Schistosoma mansoni is maintained by materials synthesized in the
underlying tegumental cell bodies. The movement or translocation
of these materials from the cell bodies where they are produced
to more superficial areas apparently is dependent upon microtubules (Bogitsh, 1977).

II. MICROTUBULES

The presence of microtubules has been observed in *S. mansoni* by
a number of investigators. Morris and Threadgold (1968) describe
microtubules in the tegumental cytoplasmic channels, while Dike
(1971) and Rifkin (1971) report the presence of these structures
in the folds of the esophageal epithelium and in the tegument of
schistosomules, respectively. Wilson and Barnes (1974) present
evidence that the cytoplasmic channels of the worm's tegument are
lined by a peripheral ring of microtubules. The position of these
microtubules on the inner surface of the plasma membranes of cell
ducts and in various spermatozoa leads many investigators to
ascribe to them the function of providing cytoskeletal support
(Halton and Dermott, 1967; Spence and Silk, 1970; Wikel and
Bogitsh, 1974; and others).

The first suggestion that microtubules may be involved in a cell's secretory process is that issued by Lacy *et al.* (1968) while investigating the secretion of insulin by the vertebrate pancreas. Since that report, secretory activities have been investigated extensively in a variety of vertebrate secretory cells. Antimicrotubular drugs have been shown to inhibit the secretions of thyroxine (Williams and Wolff, 1970), plasma protein (Redman *et al.*, 1975), hepatic, very low-density lipoproteins (LeMarchand *et al.*, 1973; Stein and Stein, 1973), proparathormone (Reaven and Reaven, 1975; Chu *et al.*, 1977), and a number of others (see Redman *et al.*, 1975, and Hoffstein *et al.*, 1977 for reviews).

The structures in *S. mansoni* which appear to be most dependent on microtubule-directed secretory activities, hence, most susceptible to the type of chemotherapeutic attack being proposed, are the tegumental cell bodies and the esophageal gland. The morphology of the surface covering, or tegument, of *S. mansoni*, and parasitic flatworms in general, has been well documented, and the details need not be elaborated herein (see Lumsden, 1975a,b for reviews). The cytoarchitecture of the surface of the organism is essentially syncytial. The underlying tegumental cell bodies are connected to the outer tegument by cytoplasmic channels which traverse two or three layers of muscle. A number of investigators consider the tegumental cell body to be the site of synthesis for materials which, in turn, maintain the tegument and attendant structures such as the glycocalyx. The secretions, usually contained in discrete membrane bound vesicles, are transported to the tegument via the cytoplasmic channels (Fig. 1).

The surface of *S. mansoni* plays a vital role in the adaptability of the organism to its environment. Not only does it determine which physiologically important materials are assimilated from the host, but it also controls the degree of immunity to hostile environmental forces generated by the host since it is against this structure that the host's defense mechanisms are directed. The esophageal gland is a modification of the tegumental cell bodies in the esophagus (Fig. 2), and it is hypothesized that granules secreted by this structure enables the organism to digest red blood cells (Bogitsh and Carter, 1977).

A range of techniques has been employed to demonstrate convincingly that the tegumental cell bodies and the esophageal gland cells are responsible for synthesis of material. In the case of the tegumental cell bodies, the secretions serve to maintain the tegument and its glycocalyx. For instance, Lumsden (1966), Oaks and Lumsden (1971), Shannon and Bogitsh (1971), and Kusel *et al.* (1975) have utilized autoradiographic procedures to show that proteins and macromolecular carbohydrates are packaged via the Golgi complexes and subsequently concentrated in the more superficial regions of the tegumental syncytium. Bogitsh (1968), Oaks and Lumsden (1971), Wilson and Barnes (1974), Stein and Lumsden

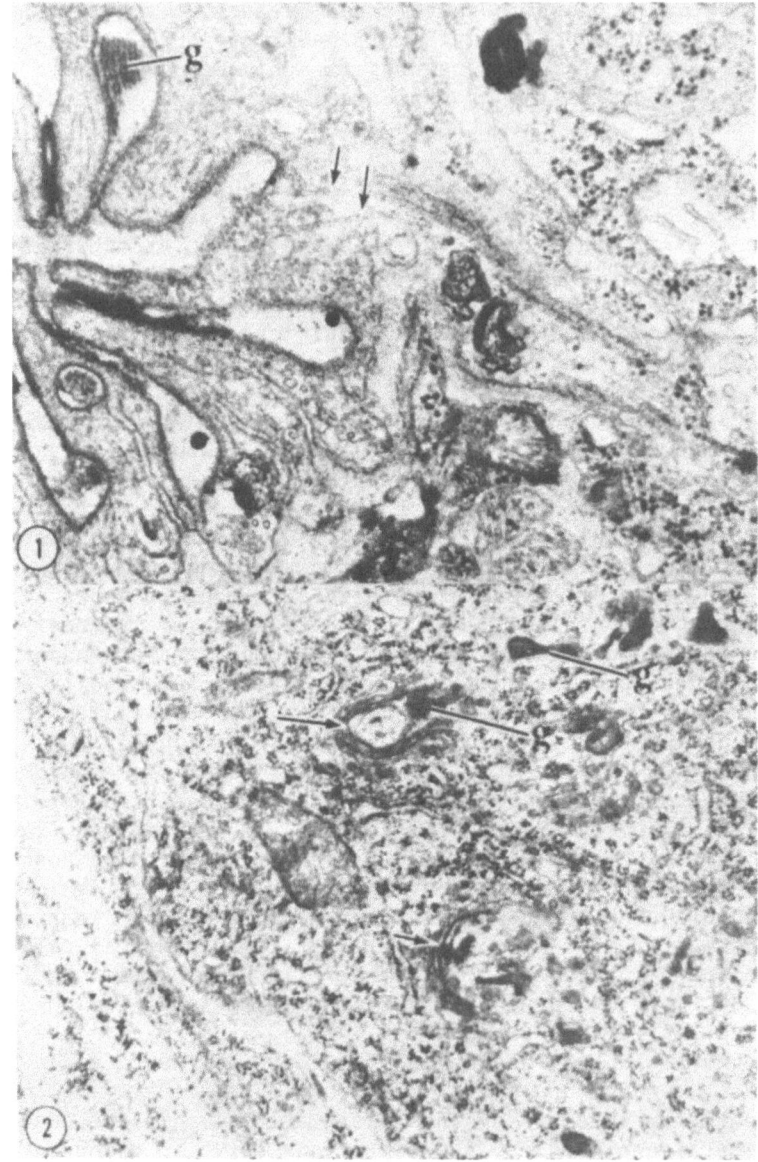

Fig. 1. Section through the posterior portion of the esophagus of
         Schistosoma mansoni showing microtubules (arrows) and part
         of a membranous secretory granules in the luman (g). X 49,400.

Fig. 2. Section through a portion of the esophageal gland of
         Schistosoma mansoni showing Golgi complexes (arrows) and
         some granules (g). X 30,600.

(1973), Dike (1971), Ernst (1975), Threadgold (1968), and others have used a variety of histo- and cytochemical techniques for further documentation of this phenomenon. Similarly, Bogitsh and Carter (1977) have demonstrated a similar sequence of intracellular events in the esophageal gland cells of *S. mansoni* during the production of a membranous secretory granule. In general, these studies and others in a variety of parasitic flatworms demonstrate the presence of secretory vesicles or granules in the cell bodies of the tegument and esophagus which display cytochemical continuity with Golgi saccules and, at times, the glycocalyx. Biochemical reports support these observations in a number of instances (Oaks and Lumsden, 1971; Lumsden, 1975b). It, therefore, becomes clear that the synthetic activities in these tissues occur along the usual intracellular pathways. The missing piece to the puzzle in *S. mansoni* was how the secretions were translocated through the cytoplasmic channels to regions nearer the surface.

III. COLCHICINE AND VINBLASTINE

In studies of vertebrate tissues following the report of Lacy *et al.* (1968), alkaloid drugs such as colchicine and the *Vinca* alkaloids, among others, have been used to depolymerize the protein tubulin of which the microtubules are constituted. When schistosomules of *S. mansoni* are exposed to colchicine (5 X 10^{-4}M, 2 hr) *in vitro*, definite morphological changes occur. Microtubules disappear from the cytoplasmic channels of the two tissues investigated, i.e., the tegument and the posterior portion of the esophagus. Concomitant with the disappearance of the microtubules, there appears a massive accumulation of secretory products in both the subtegumental cell bodies and the esophageal gland cells (Bogitsh and Carter, 1977; Bogitsh, 1977) (Figs. 3, 4). This phenomenon substantiates the report of Stein *et al.* (1974) that colchicine does not effect the synthetic machinery of the cell but inhibits the eventual discharge of the product (Fig. 3). Also, no secretory products are visible either in the tegument proper or in the esophageal lumen. Results are similar when vinblastine is substituted in the culture for colchicine. However the *Vinca* alkaloids such as vinblastine and vincristine bind to the tubulin molecule at different sites from those of colchicine (Wilson *et al.*, 1974) and causes the tubulin to polymerize into crystalline configurations (Fig. 4) with specific characteristics as described in detail by Bensch and Malawista (1969). Further, when drug-treated worms are offered red blood cells, fewer than 1% display any sign of pigment in the gut indicating a lack of hemoglobin digestion. By contrast, an average of 87% of control worms display pigment in the gut.

Fig. 3.   Section through esophageal gland of *Schistosoma mansoni*
          treated with colchicine (5 x 10^{-4}M). Note the accumu-
          lation of granules surrounding the nucleus (N). The
          gastrodermis is at the right (G). X 7,200.
          Insert: The synthetic machinery of the colchicine treated
          cell appears to be normal as evidenced by a well developed
          Golgi complex. X 59,200.

Fig. 4.   Section through the tegumental region of *Schistosoma
          mansoni* previously treated with vinblastine (5 x 10^{-4}M).
          The tegument (T) is at the left. Note the crystals
          (arrows) in the subtegumental cell body. These probably
          represent a polymerization of tubulin in a crystalline
          array characterized by hexagonal subunits (see insert).
          X 30,600.
          Insert: A higher magnification of a vinblastine-
          induced crystal. X 78,200.

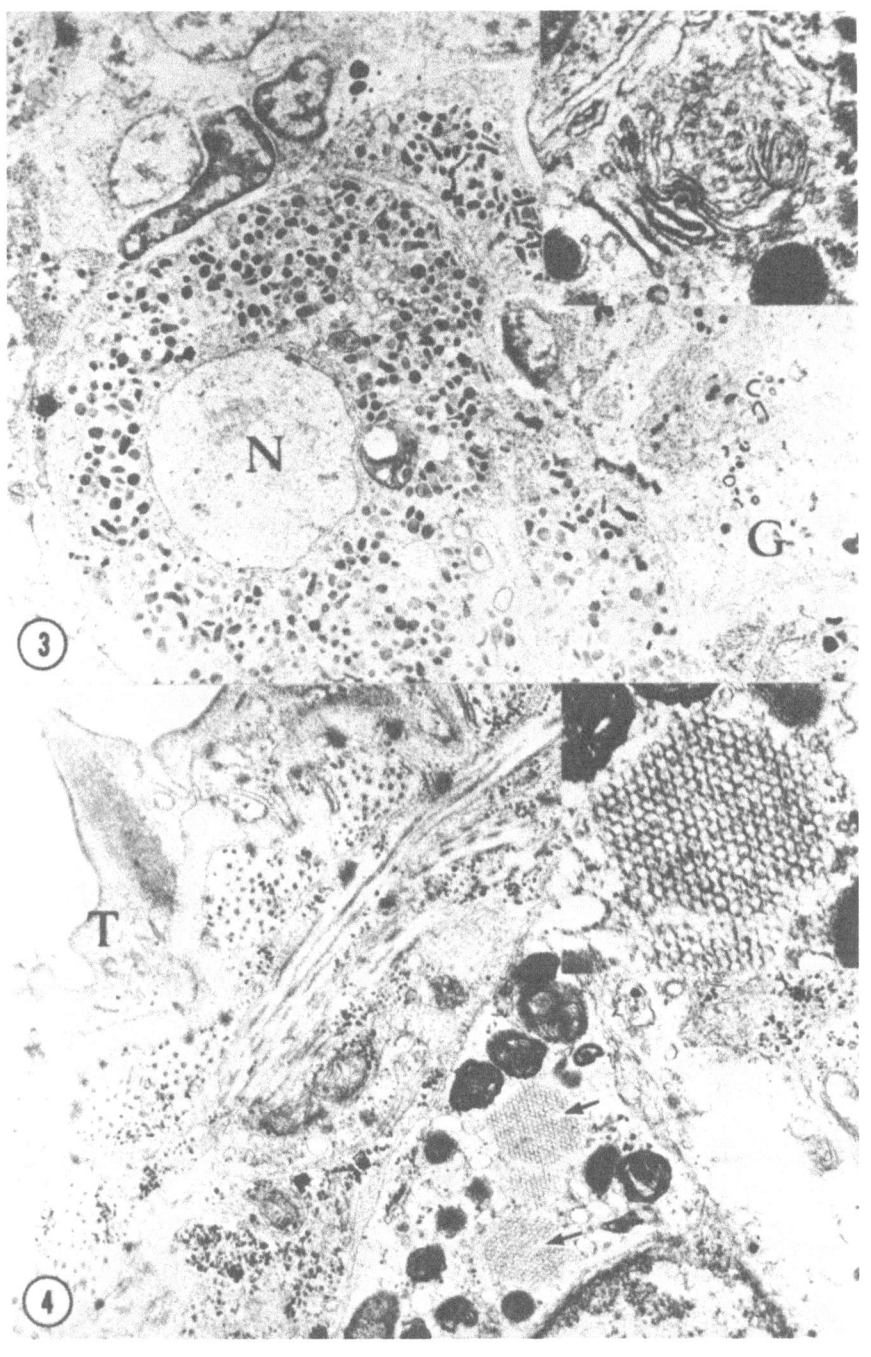

CONCLUSIONS

While further experimentation is required, it is nevertheless clear that microtubules present a promising target for chemotherapeutic attack. At least two life processes in *S. mansoni* are dependent upon microtubule-directed secretory activities, maintenance of the tegumental glycocalyx, and the production of esophageal granules (Bogitsh and Carter, 1977). In both, secretions are produced by subtegumental cell bodies such as the esophageal gland and are translocated to more superficial areas. It is the translocation of these secretions that is dependent upon microtubules. The glycocalyx is essential for protection of the organism against the host's immune mechanisms. Kusel *et al.* (1975) report that the half-life of the glycocalyx of *S. mansoni* adults is approximately 30 hr, indicating the need for continual renewal of this structure. As for the digestive system, Bogitsh and Carter (1977) and Bogitsh (1977) report that colchicine inhibits the digestion of red blood cells by schistosomules *in vitro* and that the esophageal gland is the tissue most dramatically affected by the drug. Understanding of the precise mechanism involved in such microtubule-directed secretory activities in the schistosomes, though fragmentary at present, appears to hold promise as a useful instrument in the development of new chemotherapeutic agents in the treatment of the disease. Such agents designed to interrupt and/or inhibit this secretory activity could provide an effective means of control of the causative organism and hence the disease.

REFERENCES

Bensch, K. G. and Malawista, S. E. (1969). Microtubular crystals in mammalian cells. *J. Cell Biol.*, 40, 95-107.
Bogitsh, B. J. (1968). Cytochemical and ultrastructural observations on the tegument of the trematode *Megalodiscus temperatus*. *Trans. Am. Micros. Soc.*, 87, 477-486.
Bogitsh, B. J. (1977). *Schistosoma mansoni:* colchicine and vinblastine effects on schistosomule digestive tract development *in vitro*. *Exp. Parasitol.*, 43, 180-188.
Bogitsh, B. J. and Carter, O. S. (1977). *Schistosoma mansoni:* ultrastructural studies on the esophageal secretory granules. *J. Parasitol.*, 63, 681-686.
Cheng, T. C. (1977). The control of parasites: the role of the parasite. Uptake mechanisms and metabolic interference in parasites as related to chemotherapy. *Proc. Helminth. Soc. Wash.*, 44, 2-17.
Chu, L. L. H., MacGregor, R. R., and Cohn, D. V. (1977). Energy-dependent intracellular translocation of proparathormone. *J. Cell Biol.*, 72, 1-10.

Dike, S. C. (1971). Ultrastructure of the esophageal region in *Schistosoma mansoni*. *Am. J. Trop. Med. Hyg.*, 20, 552-568.

Ernst, S. C. (1975). Biochemical and cytochemical studies of digestive-absorptive functions of esophagus, cecum, and tegument in *Schistosoma mansoni:* acid phosphatase and tracer studies. *J. Parasitol.*, 61, 633-647.

Halton, D. W. and Dermott, E. (1967). Electron microscopy of certain gland cells in two digenetic trematodes. *J. Parasitol.*, 53, 1186-1191.

Hoffstein, S., Goldstein, I. M., and Weissmann, G. (1977). Role of microtubule assembly in lysosomal enzyme secretion from human polymorphonuclear leukocytes. A reevaluation. *J. Cell Biol.*, 73, 242-256.

Kusel, J. R., Sher, A., Perez, H., Clegg, J. A., and Smithers, S. R. (1975). The use of radioactive isotopes in the study of specific schistosome membrane antigens. *In "Nuclear Techniques in Helminthology Research"* IV. p. 127-143. Intl. Atom. Energy Agency, Vienna.

Lacy, P. E., Howell, S. L., Young, D. A., and Fink, C. J. (1968). New hypothesis of insulin secretion. *Nature*, 219, 1177-1179.

LeMarchand, Y., Singh, A., Assimacopoulos-Jeannet, R., Orci, L., Rouiller, C., and Jean-Renaud, B. (1973). A role for the microtubular system in the release of very low density lipo-proteins by perfused mouse livers. *J. Biol. Chem.*, 248, 6862-6870.

Lumsden, R. D. (1966). Cytological studies on the absorptive surfaces of cestodes. II. The synthesis and intracellular transport of protein in the strobilar integument of *Hymenolepis diminuta*. *Zeit. Parasitenk.*, 28, 1-13.

Lumsden, R. D. (1975a). Surface ultrastructure and cytochemistry of parasitic helminths. *Exp. Parasitol.*, 37, 267-339.

Lumsden, R. D. (1975b) The tapeworm tegument: a model structure and function in host-parasite relationships. *Trans. Amer. Micros. Soc.*, 94, 501-507.

Morris, G. P. and Threadgold, L. T. (1968). Ultrastructure of the tegument of adult *Schistosoma mansoni*. *J. Parasitol.*, 54, 15-27.

Oaks, J. and Lumsden, R. D. (1971). Cytological studies on the absorptive surfaces of cestodes. V. Incorporation of carbohy-drate containing macromolecules into tegument membranes. *J. Parasitol.*, 57, 1256-1268.

Reaven, E. P. and Reaven, G. M. (1975). A quantitative ultra-structural study of microtubule content and secretory granule accumulation in parathyroid glands of phosphate- and colchicine-treated rats. *J. Clin. Invest.*, 56, 49-55.

Redman, C. M., Banerjee, D., Howell, K., and Palade, G. E. (1975). Colchicine inhibition of plasma protein release from rat hepatocytes. *J. Cell Biol.*, 66, 42-59.

Rifkin, E. (1971). Interaction between *Schistosoma mansoni* schistosomules and penetrated mouse skin at the ultrastructural level. *In "The Biology of Symbiosis"* p. 25-43. (T. C. Cheng, ed.). University Park Press, Baltimore, Maryland.

Senft, A. W. (1969). Considerations of schistosome physiology in the search for antibilharziasis drugs. *Ann. N. Y. Acad. Sci.* 160, 571-592.

Shannon, W. A. and Bogitsh, B. J. (1971). *Megalodiscus temperatus:* comparative radioautography of glucose $-^3$H and galactose $-^3$H incorporation. *Exp. Parasitol.*, 29, 309-319.

Spence, L. M. and Silk, M. H. (1970). Ultrastructural studies of the blood fluke - *Schistosoma mansoni*. IV. The digestive system. *S. Afr. J. Med. Sci.*, 35, 93-112.

Stein, O. and Stein, Y. (1973). Colchicine-induced inhibition of very low density lipoprotein release by rat liver *in vivo*. *Biochim. Biophys. Acta*, 306, 142-147.

Stein, O., Sanger, L., and Stein, Y. (1974). Colchicine-induced inhibition of lipoprotein and protein secretion into the serum and lack of interference with secretion of biliary phospholipids and cholesterol by rat liver *in vivo*. *J. Cell Biol.*, 62, 90-103.

Stein, P. and Lumsden, R. D. (1973). *Schistosoma mansoni:* topochemical features of cercariae, schistosomula, and adults. *Exp. Parasitol.*, 33, 499-514.

Wikel, S. K. and Bogitsh, B. J. (1974). *Schistosoma mansoni:* penetration apparatus and epidermis of the miracidium. *Exp. Parasitol.*, 36, 342-354.

Williams, J. A. and Wolff, J. (1970). A possible role of microtubules in thyroid secretion. *Proc. Natl. Acad. Sci. U.S.A.*, 67, 1901-1908.

Wilson, L., Bamburg, J. R., Mizel, S. B., Grisham, L. M., and Cresswell, K. (1974). Interaction of drugs with microtubule proteins. *Fed. Proc.*, 33, 158-166.

Wilson, R. A. and Barnes, P. E. (1974). The tegument of *Schistosoma mansoni:* observations on the formation, structure, and composition of cytoplasmic inclusions in relation to tegument function. *Parasitology*, 68, 239-258.

Developing an Invertebrate Model for Chemical Carcinogenesis: Metabolic Activation of Carcinogens

ROBERT S. ANDERSON

Sloan-Kettering Institute for Cancer Research
145 Boston Post Road
Rye, New York

I. INTRODUCTION

It has been reported occasionally that the incidence of neoplasia among invertebrate animals is lower than that seen among
mammalian species. It is unclear if this is true or if this is
a result of lack of data, or some unique feature of invertebrate
biology. Since many chemical carcinogens require metabolic
activation before they can express their biological activities,
lack of the enzymes which mediate these reactions could supply
a basis to support the theory that invertebrates are resistant
to cancer. Therefore, this laboratory initiated comparative
studies of aryl hydrocarbon hydroxylase (AHH) activity in lower
animals using benzo[a]pyrene (BP) as substrate.

Benzo[a]pyrene is a ubiquitous environmental pollutant which
is a potent carcinogen for mammalian species. However, its
biological effect on invertebrates is only poorly understood.
Benzo[a]pyrene is one of the polycyclic aromatic hydrocarbon
(PAH) oncogens which is thought to require enzymatic activation.
Before studying its effects on invertebrates, we thought it
important to see if these organisms have the capacity to
metabolize BP. This was thought to be the logical first step
in developing a model system for studying chemical carcinogenesis
in lower animals. It has been shown that BP metabolites can bind
covalently to biological macromolecules, such as DNA, while the
parent compound cannot.

II. EXPERIMENTAL ANIMALS

The southern armyworm, *Spodoptera eridania*, was the first
invertebrate examined in this laboratory for AHH activity.
Spodoptera was selected because it was known to have high mixed-
function oxidase (MFO) activity, as assayed by its ability to
convert aldrin to dieldrin (Krieger *et al.*, 1971). Also, the
invertebrate MFO literature was fairly extensive for insects,
primarily because of interest in the metabolism of insecticides.
At this time, we have worked most intensively on *Spodoptera*;
however, studies have now extended to various aquatic invertebrates.

We have been particularly interested in AHH in the oyster
Crassostrea virginica, and have also some data on several mussels,
Mytilus edulis and *Geukensia demissa*. Shellfish are known to
concentrate oncogens such as BP within their tissues (Dunn and
Stich, 1975). Marine mollusks were reported to lack AHH (Lee
et al., 1972; Payne, 1976; Vandermeulen *et al.*, 1977); however,
there are also suggestions that molluscan neoplasia can be
correlated to PAH pollution (Yevich and Barszcz, 1976; Brown *et
al.*, 1977). Furthermore, aside from possible carcinogenic effects

on themselves, the activity of molluscan AHH could mean that organisms which ingest them, including man, might be immediately exposed to carcinogenic BP derivatives.

III. ARYL HYDROCARBON HYDROXYLASE ASSAY CONDITIONS

The reaction being quantified in all invertebrate studies reported here is the *in vitro* generation of labeled BP metabolites, using invertebrate digestive gland homogenates or microsomes as an AHH source. This method is more sensitive than those which measure fluorescence.

It was essential to develop optimal conditions for demonstrating AHH in each of the invertebrate species. It was necessary to optimize temperature, pH, ionic composition, type of buffers, cofactor concentrations, etc. before the kinetics of the reaction could be determined. Probably the most often used method for measuring benzo[a]pyrene hydroxylase activity in mammals is that of Nebert and Gelboin (1968); however, these incubation conditions are inadequate for invertebrates such as oysters. For example, the NADPH concentration required for mammalian preparations is so far in excess of that required for mollusks as to be inhibitory. Also, invertebrate MFO activity is high only in the digestive gland (hepatopancreas in oysters); in mammals the enzymes are most active in the liver. Therefore, it is not surprising that other workers could show no AHH activity using the method of Nebert and Gelboin (1968) on whole bivalve homogenates.

IV. ENZYME ASSAY IN *SPODOPTERA ERIDANIA*

The 6th instar larvae were decapitated, and the gastrointestinal tracts were carefully excised and placed in ice-cold 0.05 M Tris (tris-[hydroxymethyl]aminomethane)HCl buffer, pH 7.8. The midgut region was dissected and slit open longitudinally. The midgut contents were removed and the tissues washed thoroughly with ice-cold KCl. Tissue homogenates (2 midguts/0.1 ml) were prepared in 0.15 M KCl in 0.05 M Tris-HCl buffer. A microsomal preparation was obtained by centrifugation of the midgut homogenate. The homogenates were first centrifuged at 10,886 g for 10 min in a Sorvall RC2-B automatic refrigerated centrifuge using an SS-34 rotor. The supernatant from each tissue homogenate was recentrifuged at 10,886 g, and the resulting supernatants were ultracentrifuged to obtain microsomal pellets. This centrifugation was carried out at 100,000 g using a Beckman Model L preparative ultracentrifuge with a 40 rotor. The microsomal pellets were immediately resuspended in ice-cold 0.15 M KCl in 0.05 M Tris so that 0.1 ml contained microsomes from four midguts. All midgut homogenates and/or microsomal preparations were held

in an ice bath until the enzyme assays were performed. In no case
did this period exceed 30 min. An aliquot of each homogenate
or microsome preparation was frozen for subsequent protein
determination by the method of Lowry *et al.* (1951).

One ml reaction mixtures were prepared in 10-ml Erlenmeyer
flasks, which were held on ice until the incubation stage. The
final concentrations of the components of the reaction mixture
were 5 mM NADPH (nicotinamide adenine dinucleotide phosphate,
reduced form), 3 mM KCl, 1.6 mg/ml albumin (bovine serum
albumin, fraction V) in 0.05 M Tris-HCl buffer. All the above
compounds were purchased from the Sigma Chemical Company, St.
Louis, Missouri. Each flask contained either 0.1 ml of microsomes or 0.2 ml whole midgut homogenate which represented about
3-5 mg protein. The reaction was initiated by the addition of
80 nmoles (0.1 μCi) 7,10-^{14}C-benzo[a]pyrene (BP). ^{14}C-benzo[a]-
pyrene, specific activity 201 μCi/mg, was purchased from
Amersham/Searle Corporation, Arlington Heights, Illinois. A
working stock containing 4 μCi ^{14}C-BP/ml acetone was made by
diluting the original stock with unlabeled BP. The desired
amount of ^{14}C-BP was introduced to the reaction mixture by
delivering 0.025 ml of working stock to each flask.

The experimental flasks were incubated 10 min at 30°C in an
Eberbach shaking water bath, after which time the reaction was
stopped by the addition of 1 ml cold acetone. The reaction in
the control (T_0) flasks was stopped immediately after the
addition of ^{14}C-BP. Once the reactions were stopped both T_{10} and
T_0 samples were subjected to the same methodological protocol.
The reaction mixtures were extracted with 2 ml hexane for 15 min
at 30°C on the shaking water bath. A 1 ml aliquot of the hexane
layer was removed and vigorously mixed with an equal volume of 1
N NaOH; then a 0.5 ml sample of the NaOH extract was placed in a
scintillation vial. The aqueous layer was extracted again with
2 ml hexane. The hexane layer was removed entirely, and a 0.5 ml
aliquot of the aqueous layer was pipetted into a scintillation
vial. The protein in this aliquot was solubilized by overnight
incubation with 0.25 ml Protosol (New England Nuclear, Boston,
Massachusetts) at room temperature. Finally, the radioactivity
present in the NaOH and aqueous aliquots was quantified using a
Packard Model 3255 liquid scintillation spectometer. The extrac-
tion method was based on that of Abramson and Hutton (1975) with
modifications by L. M. Anderson (pers. comm.).

The amount of radioactivity present in the total NaOH and
aqueous layers from a given sample was calculated taking into
consideration quenching, machine efficiency, dilution, and par-
titioning. The nmoles of ^{14}C-BP metabolic products generated
during the course of the experiment was determined using the
specific activity. The activity of the enzyme benzo[a]pyrene
hydroxylase was expressed as nmoles ^{14}C-BP metabolized per mg
protein per 10 min.

V. ENZYME ASSAY IN *CRASSOSTREA VIRGINICA*

The hepatopancreas was excised and homogenized in ice-cold 0.839 M sucrose. Homogenates and/or microsomes were prepared as described in the preceding section.

One ml reaction mixtures were prepared in 10-ml Erlenmeyer flasks, which were held on ice until the incubation stage. The final concentrations of the components of the reaction mixture were 0.1 mM NADPH, 5 mM Mg SO_4, and 1.6 mg/ml bovine serum albumin in 0.4 M HEPES (N-2-hydroxyethylpiperazine-N'-2-ethane sulfonic acid) buffer, pH 7.8. The addition of ^{14}C-BP was carried out as previously described. Incubation temperature was 30°C; reactions in control tubes were stopped at T_o, experimentals at T_{60}. Aqueous and NaOH fractions were prepared by the standard method.

VI. CHARACTERISTICS OF *SPODOPTERA* AHH

Typical of MFO activity, *Spodoptera* AHH required NADPH and oxygen. Enzymatic activity increased proportionally with NADPH concentrations <5 mM, but was unaffected by, or inhibited by, higher concentrations. Various concentrations of NADH failed to alter the rate of reaction; it could appear that NADPH is the preferred electron donor. Oxygen dependency of the reaction was shown by N_2 inhibition. The metabolism of BP was markedly reduced by bubbling N_2 through the reaction mixture.

In vitro metabolism of BP by *Spodoptera* AHH was maximal from 30°-40°C, but undetectable at 56°C. This indicates that the enzyme is thermostable at temperatures in excess of those likely to exist *in vivo*. In working with mammalian AHH it is often useful to protect -SH groups by the addition of DL-dithiothreitol (DDT) to the reaction mixture. However, in our system DDT had no significant effect. The pH optimum was shown to fall in the range pH 7.6-8.0. Half-maximal activity was obtained at about pH 7.2 and 8.4, with activity falling off rapidly as the pH varied further from 7.8.

The production of BP metabolites continues for at least 60 min in this *in vitro* system; both water- and NaOH-soluble products can be detected after 2 min. The rate of reaction is linear for the first 20 min, but diminishes thereafter. AHH activity is expressed as nmoles BP metabolized/mg protein/10 min; 10 min is on the linear portion of the curve. The amount of metabolites generated in 10 min increases with AHH concentration, as measured by homogenate or microsome protein content, under about 3.5 mg/ml. In these studies about half this concentration was used.

The midgut, or stomach, appeared to be the only tissue rich in BP hydroxylase. No activity was present in head capsule or samples of integument, slight activity was seen in the crop,

anterior intestine, rectum and Malpighian tubules. This is
generally the case for other MFO activities in insects, which are
usually associated with the midgut, Malpighian tubules, and/or
fat body. The subcellular distribution of AHH was determined
by assaying activity in various cell fractions. The pellet
containing nuclei and cell debris had about 25% of the AHH
activity of whole midgut homogenate; however, the microsomal
fraction showed about 400% of the homogenate activity. Therefore,
Spodoptera BP hydroxylase, like other invertebrates and vertebrate MFO activities, is located primarily in the cellular
endoplasmic reticulum.

As mentioned previously, the head capsule had no AHH activity.
In fact, this structure contained a water-extractable factor
which is a potent natural inhibitor of AHH. It had been
established that the insect eye pigment xanthommatin is an
effective aldrin epoxidase inhibitor in the housefly (Schonbrod
and Terriere, 1971). Early in our studies it was clear that
little or no AHH activity could be shown in midgut preparations
which had not been washed free of their contents. The addition
of buffered homogenate of *Spodoptera* gut contents to microsomes
effectively abolished all enzyme activity. Similar inhibitors,
which are probably proteolytic enzymes, have been described by
other investigators (Krieger and Wilkinson, 1970; Krieger and
Lee, 1973; Brattsten and Wilkinson, 1973).

The presence of these endogenous AHH inhibitors emphasizes the
necessity to work with clean, homogenous tissue samples and
explains why it may be difficult to detect enzymatic activities
in homogenates of whole organisms. Both inhibitors are probably
agents produced by tissue disruption which have little influence
on AHH *in vivo*.

Krieger and Wilkinson (1969) reported marked changes in epoxi-
dase-specific activity during larval development of *Spodoptera*.
We have shown that AHH activity also was quite low in young
larvae and that this level gradually increased with maturity.
This effect was not caused by enzyme induction, since the diet
remained unchanged throughout the study, and the animals were
never exposed to xenobiotics of any kind. Before pupating, the
larvae entered a wandering stage during which time they did not
feed. AHH specific activity in these prepupal larvae was very
low.

The reactions carried out by both mammalian and invertebrate
MFO are mediated by a cytochrome, such as cytochrome P450, which
is present in insect preparations (Ray, 1967; Gil *et al.*, 1974).
Carbon monoxide is a strong cytochrome inhibitor; the AHH activity
of *Spodoptera* midgut homogenates was reduced by more than 90%
when treated with CO. Carbon monoxide sensitivity of insectan
AHH is another parameter indicating that this enzyme shares all
of the important characteristics of typical microsomal mixed-
function oxidases.

VII. COMPARISON OF BP HYDROXYLASE ACTIVITY IN *SPODOPTERA* AND MOUSE

Mouse results were obtained using conditions optimal for that organism; the incubation mixture used was essentially that of Abramson and Hutton (1975). The only modifications were the inclusion in the 1 ml reaction mixtures of 0.5 µmoles NADP and 0.4 µmoles NADH.

The BP hydroxylase activity in *Spodoptera* preparations was somewhat variable. This might be explained on the basis of genetic differences, age differences, or merely natural variation. The animals are of at least similar genetic composition since they come from an inbred colony which has been maintained for more than a decade. Differences in activity caused by age variations were minimized by using last instar, mature larvae for the assays.

The mean total BP hydroxylase activity in midgut homogenates from untreated mature *Spodoptera* larvae was 0.398 ± 0.150 nmoles BP metabolized/mg protein/10 min (n = 15). This represents the sum of the phenolic BP metabolites contained primarily in the NaOH extract, which average 0.229 ± 0.131 nmoles/mg protein/10 min, plus the hydroxylated BP and conjugated metabolies in the aqueous layer (0.169 ± 0.044 nmoles/mg protein/10 min).

The data for uninduced mouse hepatic homogenates are as follows: Total BP metabolism is 2.12 nmoles/mg protein/10 min in C57BL/6 ♂ (n = 8) and 2.37 nmoles in Camm Swiss-Webster ♂ (n = 8). Approximately equal amounts of NaOH- and water-soluble metabolites were generated by mouse liver homogenates.

It would appear that the base line activity of benzo[a]pyrene hydroxylase in laboratory mice is about 5-6 times that of *Spodoptera*. However, one cannot conclude that insects have generally less active MFO systems. For example, Krieger and Wilkinson (1969) found that the specific activity of *Spodoptera* epoxidase surpassed the most active mammalian liver preparation (data from 7 species reported) by a factor of almost 10.

VIII. INDUCTION OF *SPODOPTERA* AHH BY POLYCHLORINATED BIPHENYLS

Polychlorinated biphenyls (PCBs), which are ubiquitous and stable environmental pollutants, are potent inducers of hepatic AHH and other MFO activities. These studies were designed to quantify the AHH inductive effect of various common commerical PCB mixtures on *Spodoptera*. Agents which alter MFO activity can have profound effects on important chemical transformations including metabolism of xenobiotics (such as drugs, insecticides, and chemical oncogens) and steroid hormones. In many species of insects chlorinated hydrocarbons, especially insecticides, have been shown to induce MFO (Morello, 1964; Gil *et al.*, 1968; Agosin *et al.*, 1969; Plapp and Cassida, 1970; Khan and Matsumura,

1972);however, other investigators have failed to show this
effect (Chakraborty and Smith, 1967; Meksongsee *et al.*, 1967;
Oppenoorth and Houx, 1968; Perry and Buckner, 1970).

Commercial PCB mixtures were generously provided by the
Monsanto Company, St. Louis, Missouri. These included Capacitor
21, Aroclor 1016, Aroclor 1242, and Aroclor 1254, which contain
21, 41, 42, and 54% chlorine by weight, respectively. Several
PCB-containing microscope immersion oils were also used; these
were Cargille types A and B. In all cases, weighed amounts of
undiluted PCBs were applied to a narrow area of the dorsal
surface of the larvae. The amounts of PCB administered were not
lethal over the course of the study. No attempt was made to
quantify the amount of PCB actually absorbed by the organisms;
however, it is unlikely that much of the administered dose was
lost by chance contact with housing material or other larvae.

In a typical experiment at least four groups of animals of the
same age received identical laboratory exposure to a give PCB.
At the time of AHH assay, the midgut regions of all the animals
in a particular experimental group were pooled and homogenates
or microsomes prepared. The variation between the data from all
groups was expressed statistically. Untreated control groups
were analyzed in a similar fashion.

Production of total ^{14}C-BP metabolites was followed during the
first 60 min incubation of midgut homogenate from animals treated
24 hr previously with 1.3 mg Aroclor 1254. The rate of reaction
for PCB-induced homogenate, as well as uninduced homogenate, is
linear with time for 10 min. Therefore, all assays reported
here were terminated at this time. The rate of reaction was
more rapid for induced preparations than for controls; by 20 min
the controls produced about 70% of the 60 min total metabolites;
in the same time the Aroclor-induced samples produced about 95%.
However, induction had little effect on the relative proportion
of NaOH-soluble and water-soluble BP derivatives produced.

The activity of AHH induced by various PCBs was measured 24 hr
after topical administration. The larvae used in all induction
studies were fully mature and weighed 690 ± 105 mg. Known
amounts of Aroclor were applied to the animals over the range
0.1-5.0 mg/larvae. The 24 hr response to all PCBs tested is
maximal between doses of about 1-2 mg and shows no significant
fluctuation over this range. Doses lower than 0.5 mg usually
produced a less than maximal response, as did doses in excess
of 3 mg. The decreased induction at the higher doses probably
reflected PCB toxicity produced after 24 hr. We applied 1.0-1.5
mg PCB to each animal in our subsequent experimental protocols;
animals so treated showed no overt toxic reactions.

Dermal application of commonly used PCB-containing microscope
immersion oils also induced AHH activity in *Spodoptera*. Doses
of 0.5-1.0 mg of Cargille immersion oil types A and B (Cargille
Laboratories, Cedar Grove, New Jersey) induced a 2-4-fold increase

in benzo[a]pyrene hydroxylase activity in 24 hr. Oils A and B contain the following constituents: mineral oil, 47 and 29; polybutene, 19 and 43; and Aroclor 1254, 34 and 28 percent by weight, respectively. Significant increases in hepatic benzo[a]pyrene hydroxylation were also measured in rats after skin application of as little as 1 μl of microscope immersion oil, known to contain PCBs (Bickers *et al.*, 1975).

IX. EFFECT OF PCB CHLORINE CONTENT ON AHH INDUCTION IN *SPODOPTERA*

Several studies suggest that, at least in some species, the MFO inductive capacity of PCBs may be directly proportional to their chlorine content. Chen and DuBois (1973) studied the hepatic microsomal enzyme-inducing effects of Aroclor 1221, 1254, and 1260 in young rats. Enzyme induction persisted for at least 13 weeks and was proportional to chlorine content for phosphorothio-ate detoxification and N-demethylation, using aminopyrine. A direct relationship between PCB chlorine content and induction of phenobarbital hydroxylation and nitroreductase activity has also been established for rats (Litterst *et al.*, 1972). A similar conclusion was reached by Rhee and Plapp (1973) who reported that the amount of aldrin epoxidase induction in houseflies was directly proportional to Aroclor chlorine content.

The PCBs tested on *Spodoptera* include Capacitor 21 and Aroclors 1016, 1242, and 1254, containing 21, 41, 42, and 54% chlorine, respectively. Aroclor 1254 produced a mean 24-hr induction of about four times the control level. The other PCBs produced a significantly higher AHH level in the same period; Capacitor 21 and Aroclor 1016 about a 9-fold increase, and Aroclor 1242 about a 12-fold increase. The data indicate that AHH induction in *Spodoptera* is not directly proportional to the percentage of chlorine in topically applied PCBs.

Aroclors are impure mixtures of various PCB congeners and isomers. Pure PCB isomers can be categorized into two distinct types of inducers; commerical PCB have characteristics of both groups (Goldstein *et al.*, 1977). Some isomers induced AHH, while others do so only weakly. The inductive capacity of PCB isomers is determined by distribution of chlorinated positions; their toxic and pharmacological properties cannot be predicted on the basis of degree of chlorination alone.

X. AHH INDUCTION AS A FUNCTION OF TIME

Maximum induction for both Aroclor 1242 and 1254 occurred at 24 hr after a single skin application. Subsequent to 24 hr, AHH activity decreased very rapidly for Aroclor 1242; however, 1254-induced activity persisted at near maximum level for at least 4 days.

The response to a single exposure to PCB, particularly Aroclor 1242, resulted in marked AHH induction which is transitory in duration. This indicates that detoxifying enzymes can be rapidly induced following exposure of an insect to environmental pollutants and that this effect can be quickly reversed when the animal is no longer exposed to the noxious agents.

In mammals, PCB treatment is accompanied by liver hypertrophy attributed to proliferation of smooth endoplasmic reticulum (SER) in the hepatic parenchymal cells. As a result of SER proliferation, the protein concentration in liver homogenates and microsomes may be elevated (Allen and Abrahamson, 1973). The increase in protein content is greatest after 2-4 weeks; however, maximal MFO activity occurs in 3-4 days. This suggests several events in enzyme induction by PCBs. Induction of MFO precedes membrane proliferation, whereas the induction of metabolite-conjugating enzymes, such as UDP-glucuronlytransferase, parallels SER proliferation (Grote *et al.*, 1975).

In no case were significant changes in *Spodoptera* homogenate or microsomal protein concentration detected during 48 hr after PCB treatment; AHH activity was maximal at 24 hr. This suggests that in insects, as in mammals, PCB-induced MFO induction results from specific gene activation, rather than from a general production of SER.

XI. AHH ACTIVITY IN OYSTERS AND MUSSELS

The differences in the composition of the reaction mixtures used to demonstrate AHH in oysters have already been mentioned. The most important difference was the low NADPH requirement for maximal oyster AHH activity; this was 1/50 of the cofactor concentration of *Spodoptera*. We also found that the presence of an NADPH generating system, composed of NADP, glucose-6-phosphate and glucose-6-phosphate dehydrogenase, did not enhance the reaction. As was the case for *Spodoptera*, NADH could not substitute for NADPH as an electron donor in the reaction. The pH optimum was about 7.6, but again the enzyme showed maximal activity over a range of about 0.3 pH units on either side of this value. *Crassostrea* AHH activity was inhibited by bubbling either N_2 of CO through the reaction mixtures, indicating that the reaction required O_2 and was mediated by a cytochrome. Similar to the enzyme in *Spodoptera*, AHH activity in molluscan preparations was vigorous at temperatures up to 40°C, suggesting thermal stability at temperatures in excess of those encountered *in vivo*. However, both hepatopancreas homogenates and microsomes metabolized BP at 15°C *in vitro*, a more physiological temperature. Molluscan preparations showed little or no AHH activity at 1°C or at 56°C.

In this system BP metabolites continued to be produced at a constant rate for at least 60 min, as opposed to about 10-15 min for *Spodoptera*. The rate of reaction was slower, with comparatively large amounts of metabolites still being produced between 60-120 min. Reactions were routinely stopped at 60 min, a point still on the linear portion of the curve where metabolite production was high.

Preliminary data indicate that *Crassostrea* AHH can be induced by PCBs. A 2-3-fold increase in AHH activity was measured 48 hr after exposure of oysters to 0.05 ml of Aroclor 1016 or 1221. The PCBs were injected into mantle cavity of oysters held dry at 4°C; these animals reamined closed, but in good health, during the time of exposure.

CONCLUSION

We have shown that certain insects and marine bivalve molluscs, as in the case for higher animals, possess mixed-function oxidase (MFO) which metabolizes benzo[a]pyrene (BP). BP is an environmental pollutant which, especially after metabolic activation, is a potent mammalian carcinogen. Chemically-induced neoplasia has not as yet been produced in these invertebrates; however, further investigations concerning BP oncogenesis are appropriate in light of their ability to metabolize the compound.

The specific activity of invertebrate aryl hydrocarbon hydroxylase (AHH) is lower than that of the comparable mammalian enzyme system. However, invertebrate AHH shares many characteristics typical of the MFO of higher animals including NADPH and O_2 dependency, association with the microsomal fraction, CO sensitivity, and inducibility by xenobiotics such as polychlorinated biphenyls. The major physiological role of MFO in all animals is probably biotransformation leading to the detoxification of compounds including toxic agents and environmental pollutants. In addition to nonreactive forms, highly mutagenic and carcinogenic BP metabolites may be produced during metabolism. Although we can measure *in vitro* BP metabolism by these invertebrates, the exact identity of the metabolites produced has not been established. This identification is necessary since only certain metabolites are capable of binding covalently to biological macromolecules such as DNA, RNA, and proteins. The final expression of BP-induced neoplasia may be determined by the balance between activation and detoxification, as well as target tissue sensitivity.

ACKNOWLEDGMENTS

This study was supported by grants from the Whitewall Foundation, R804435 from the Environmental Protection Agency and CA-08748

from the National Cancer Institute. The technical assistance of
Justine McLoughlin and Molly Cook is appreciated.

REFERENCES

Abramson, R. K. and Hutton, J. J. (1975). Effects of cigarette
    smoking on aryl hydrocarbon hydroxylase activity in lungs
    and tissues of inbred mice. *Cancer Res.*, 35, 23-29.
Agosin, M., Scaramelli, N., Gil, L. and Letelier, M. E. (1969).
    Some properties of microsomal system metabolizing DDT in
    Triatoma infestans. *Comp. Biochem. Physiol.*, 29, 785-793.
Allen, J. R. and Abrahamson, L. J. (1973). Morphological and
    biochemical changes in the livers of rats fed polychlorinated
    biphenyls. *Arch. Environ. Contam. Toxicol.*, 1, 265-280.
Bickers, D. R., Eiseman, J., Kappas, A. and Alvares, A. P. (1975).
    Microscope immersion oils: Effects of skin application on
    cutaneous and hepatic drug-metabolizing enzymes. *Biochem.
    Pharmacol.*, 24, 779-783.
Brattsten, L. B. and Wilkinson, C. F. (1973). A microsomal
    enzyme inhibitor in the gut contents of the house cricket
    (*Acheta domesticus*). *Comp. Biochem. Physiol.*, 45B, 59-70.
Brown, R. S., Wolke, R. E., Brown, C. W. and Saila, S. B.
    (1977). Hydrocarbon pollution and the prevalence of neoplasia
    in feral soft-shell clams, *Mya arenaria*. Proc. Symp.
    Environmental Pollutants, University of Connecticut, Storrs,
    Conn. In press.
Chakraborty, J. and Smith, J. N. (1967). Enzymic oxidation of
    some alkyl-benzenes in insects and vertebrates. *Biochem.
    J.*, 102, 498-503.
Chen, T. S. and DuBois, K. P. (1973). Studies on the enzyme
    inducing effect of polychlorinated biphenyls. *Toxicol.
    Appl. Pharmacol.*, 26, 504-512.
Dunn, B. P. and Stich, H. F. (1975). The use of mussels in
    estimating benzo[a]pyrene contamination of the marine
    environment. *Proc. Soc. Exp. Biol. Med.*, 150, 49-51.
Gil, D. L., Rose, H. A., Yang, R. S. H., Young, R. G. and
    Wilkinson, C. F. (1974). Enzyme induction by phenobarbital
    in the Madagascar cockroach, *Gromphadorhina portentosa*.
    Comp. Biochem. Physiol., 47B, 657-662.
Gil, L., Fine, B. C., Dinamarca, M. L., Balazs, I., Busvine, J.R.
    and Agosin, M. (1968). Biochemical studies on insecticide
    resistance in *Musca domestica*. *Entomol Exp. Appl.*, 11, 15-29.
Goldstein, J. C., Hickman, P., Bergman, H., McKinney, J. D. and
    Walker, M. P. (1977). Separation of pure polychlorinated
    biphenyl isomers into two types of inducers on the basis of
    induction of cytochrome P-450 or P-448. *Chem.-Biol.
    Interact.*, 17, 69-87.

Grote, W., Schmoldt, A. and Dammann, H. G. (1975). The metabolism of foreign compounds in rats after treatment with polychlorinated biphenyls (PCBs). *Biochem. Pharmacol.*, 24, 1121-1125.

Khan, M. A. Q. and Matsumura, F. (1972). Induction of mixed-function oxidase and protein synthesis by DDT and dieldrin in German and American cockroaches. *Pestic. Biochem. Physiol.*, 2, 236-243.

Krieger, R. I., Feeny, P. P. and Wilkinson, C. F. (1971). Detoxication enzymes in the guts of caterpillars: An evolutionary answer to plant defenses? *Science* 172, 579-581.

Krieger, R. I. and Lee, P. W. (1973). Properties of the aldrin epoxidase system in the gut and fat body of a caddisfly larva. *J. Econ. Entomol.*, 66, 1-6.

Krieger, R. I. and Wilkinson, C. F. (1969). Microsomal mixed-function oxidases in insects. I. Localization and properties of an enzyme system effecting aldrin epoxidation in larvae of the Southern Armyworm *(Prodenia eridania)*. *Biochem. Pharmacol.*, 18, 1403-1415.

Krieger, R. I. and Wilkinson, C. F. (1970). An endogenous inhibitor of microsomal mixed-function oxidase in homogenates of the Southern Armyworm *(Prodenia eridania)*. *Biochem. J.*, 116, 781-789.

Lee, R. F., Sauerheber, R. and Dobbs, G. H. (1972). Uptake, metabolism and discharge of polycyclic aromatic hydrocarbons by marine fish. *Marine Biol.*, 17, 201-208.

Lowry, O. H., Rosebrough, N. J., Farr, A. L. and Randall, R. J. (1951). Protein measurement with Folin phenol reagent. *J. Biol. Chem.* 193, 265-275.

Meksongsee, B., Yang, R. S. and Guthrie, F. E. (1967). Effects of inhibitors and inducers of microsomal enzymes on the toxicity of carbamate insecticides to mice and insects. *J. Econ. Entomol.*, 60, 1469-1471.

Morello, A. (1964). Role of DDT-hydroxylation in resistance. *Nature* 203, 785-786.

Nebert, D. W. and Gelboin, H. V. (1968). Substrate-inducible microsomal aryl hydroxylase in mammalian cell culture. *J. Biol. Chem.* 243, 6242-6249.

Oppenoorth, F. J. and Houx, N. W. H. (1968). DDT resistance in the housefly caused by microsomal degradation. *Entomol. Exp. Appl.* 11, 81-93.

Payne, J. F. (1976). Field evaluation of benzopyrene hydroxylase induction as a monitor for marine petroleum pollution. *Science* 191, 945-946.

Perry, A. S. and Buckner, A. J. (1970). Microsomal cytochrome P-450 in resistant and susceptible houseflies. *Life Sci.*, 9, 335-350.

Plapp, F. W., Jr. and Casida, J. E. (1970). Induction by DDT
    and dieldrin of insecticide metabolism by housefly enzymes.
    J. Econ. Entomol., 63, 1091-1092.
Ray, J. W. (1967). The epoxidation of aldrin by housefly micro-
    somes and its inhibition by carbon monoxide. *Biochem.
    Pharmacol.*, 16, 99-107.
Rhee, K. S. and Plapp, F. W. Jr. (1973). Polychlorinated bi-
    phenyls (PCBs) as inducers of microsomal enzyme activity
    in the housefly. *Arch. Environ. Contam. Toxicol.*, 1, 182-192.
Schonbrod, R. D. and Terriere, L. C. (1971). Eye pigments as
    inhibitors of microsomal aldrin epoxidase in the housefly.
    J. Econ. Entomol., 64, 44-45.
Vandermeulen, J. H., Keizer, P. D. and Penrose, W. R. (1977).
    Persistance of non-alkane components of bunker C oil in
    beach sediments of Chedabucto Bay, and their lack of metabolism
    by molluscs. Proc. 1977 Oil Spill Conference. *American
    Petroleum Institute publication* No. 4284, pp. 469-473.
Yevich, P. P. and Barszcz, C. (1976). Gonadal and hemato-
    poietic neoplasms in *Mya arenaria*. *Mar. Fish. Rev.*, 38, 42-43.

Alterations in Lipid Metabolism of Molluscs Due to Dietary Changes

GEORGE P. HOSKIN

Department of Surgery
Downstate Medical Center
State University of New York
Brooklyn, New York

I. INTRODUCTION

There have been few studies dealing strictly with the effect
of diet on molluscan lipid metabolism. Considerably more information on molluscan lipid anabolism is available. Much of this
information can be related to molluscan nutrition and so to the
usefulness of molluscs as models for biomedical research.

In this paper appropriate comparisons between molluscs and
vertebrates are made. In accord with the theme of using molluscs
as models some general recommendations based on their biology
are offered.

II. THE MOLLUSC AS A MODEL

A. *GENERAL DESCRIPTION*

Molluscs are nonsegmented, schizocoelomate invertebrates.
The phylum is second in number of species only to the Arthropods.
There are seven classes: Monoplacophora (e.g. *Neoplilina)*, Aplaco-
phora (=solenogasters), Polyplacophora ("chitins"), Scaphopods
("tusk shells"), Bivalvia (=pelecypods), Gastropoda (snails), and
Cephalopoda (e.g. squid and octopus). Some authors combine the
Aplacophora and Polyplacophora into the single class Amphinura.
Terrestrial, aquatic, marine and parasitic species are represented.

B. *NUTRIENT ACQUISITION*

Molluscs acquire most of their nutrients via a mouth. Direct
tegumental absorption of dissolved fatty acids has also been
demonstrated (deMoreno *et al.*, 1976b). Natural sea water contains from 2.3 to 20.5 µg/l sterols in the dissolved and suspended
matter (Kanazawa and Teshima, 1972), 10-40 µg/l fatty acids
(Slowey *et al.*, 1962; Treguer *et al.*, 1972) and also lipid-
containing detritus (Williams, 1965). Thus in experiments under
presumed starvation conditions, if filtered natural sea water
is used, nutrients including sterols and fatty acids may be
available to the molluscs.

Bacteria present in the aquatic medium may also provide an
undefined nutrient source for filter feeding molluscs (Imai and
Hatanaka, 1949; Hidu and Tubiash, 1963).

Endosymbiotic organisms (Taylor, 1973a,b) or endosymbiotic
organelles (Taylor, 1971) are present in some molluscs and provide
nutrients to their hosts. Recently, Kremer and Schmitz (1976)
provided evidence that the opisthobranch *Hermaea bifida* harbors
intracellular plastids obtained by feeding on the red alga
Griffithsia flosculosa. When the sea slugs were held in water
containing $^{14}HCO_3^-$, the CO_2 was fixed by the plastids. Labeled
compounds, including Kreb's cycle intermediates, amino acids,

sugars, and lipids, appeared in the plastids and in the tissues of the host. The pathways of synthesis have not yet been followed. Whether the specimens obtained complex molecules including lipids, or simple precursors for their own synthetic pathways was not established.

Field-collected molluscs may vary in lipid content and may be metabolically abnormal since they are frequently hosts to protozoan and metazoan parasites. Parasites alter the lipid composition of the host (Cheng, 1965; Lunetta and Vernberg, 1971; McManus *et al.*, 1975; Porter and Gamble, 1973; Williams, 1969). The serum lipid composition of *Nassarius obsoletus* is affected by the presence of rediae of *Himasthla quissentensis* (Table 1). Specifically, the ratio of serum sterol to serum free fatty acids is increased due to a relative reduction of serum sterols.

C. *LABORATORY CULTURE*

Virtually all available data on molluscan physiology are from field-collected specimens feeding on naturally available food or from laboratory specimens fed phytoplankton, higher plant material, or sodium alginate (Standen, 1951; Malek and Cheng, 1974). Culture methods are described in Smith and Chanley (1975).

Terrestrial pulmonates grow and reproduce on restricted laboratory diets (Runham, 1975). *Arion ater* can complete its life cycle when fed only lettuce (Stern, 1970) and lettuce is apparently the only food source required by *Biomphalaria glabrata* as reported in several physiological studies (Lee and Cheng, 1971a,b, 1972; Cheng and Lee, 1971).

It should be noted that although the total lipid content of lettuce (*Lattica sativa*) is only 0.2% (Sherman, 1941), the predominate fatty acids are linoleic ($C_{18:2}$) and linolenic ($C_{18:3}$) (Oudejans and van der Horst, 1974). These are essential fatty acids (EFA) in vertebrates. Recent reviews of nutritional requirements of bivalves in culture (Epifanio, 1976; Ukeles, 1976; Koganezawa, 1976) indicate the relative lack of interest in the role of lipids in molluscan diets. While many workers have considered the total carbohydrate, protein-amino acid composition or at least the total nitrogen or caloric content of diets, similar considerations of lipids have not been made.

D. *HEMOLYMPH LIPIDS*

A vast amount of data has been gathered on mammalian serum lipids and factors influencing serum lipid composition (Harper, 1967; Orten and Newhaus, 1975). Total serum lipid levels of molluscs are low compared with those of mammalian sera. Bayne (1973a), using the sulfophospho-vanillan reaction, obtained total lipid values for *Mytilus edulis* (Bivalvia) pericardial fluid of 20 mg/100 ml during starvation and 80 mg/100 ml during periods

Table 1. Serum sterol and free fatty acid content of specimens of
          Nassarius obsoletus (Gastropoda) compared with specimens
          infected with rediae of *Himasthla quissetensis* (Trema-
          toda). Values represent peak areas ± S.D. from densi-
          tometry of thin layer chromatograms prepared from snail
          serum (100 µl).

	Uninfected (10)	Infected (9)
Sterol	392.3 ± 181	288.2 ± 128
Free fatty acid	199.4 ± 124	177.2 ± 80
Ratios FFA/Sterol for individual specimens	0.50 ± 0.13	0.65 ± 0.15

of feeding. Allen (1977), using TLC densitometry, reported total plasma lipid levels of approximately 4 mg/100 ml from specimens of *Cryptochiton (=Amicula) stelleri* (Polyplacophora) during the summer. Lawrence (1965) had obtained values of 15 mg/100 ml for this species using gravimetric methods. Hoskin and Hoskin (1977), using enzymatic and TLC procedures, obtained values of 46 to 65 mg/100 ml from *Mercenaria mercenaria* (Bivalvia) serum.

Lipid composition of molluscan serum is very different from mammalian serum. Sterols are the predominant serum lipids of *N. obsoleta* (Fig. 1) and *M. mercenaria* (Fig. 2). The predominant serum lipid of *C. steleri* is sterol plus phospholipid, which together account for 75% of the total lipid. Sterol accounts for 80% of the total serum lipid of *M. mercenaria*, and free fatty acids account for 7% and 6.4% of *C. steleri* and *M. mercenaria* serum lipid, respectively. Apparently in both species the principal transport form of lipid is free fatty acid. Hemocytes also may be important in lipid transport in these species.

The diet of marine bivalves is high in polyunsaturated fatty acids (Ackman *et al.*, 1974), a condition which in mammals leads to lowered serum cholesterol (Nestel *et al.*, 1975; McGandy and Hegsted, 1975). It should be noted, however, that in fish with diets rich in polyunsaturated fatty acids the absolute serum cholesterol levels are 2-6 times those of mammals (Larsson and Fänge, 1977). Since approximately 75% of the serum cholesterol of mammals (Goodman, 1965; Harper, 1967) and 62-88% of the serum cholesterol of fish (Larsson and Fänge, 1977) is in the esterified form compared with 80% in *M. mercenaria* in the unesterified form, it seems probable that the role of cholesterol esters in fatty acid transport is of greater importance in vertebrates than molluscs. The physiological significance of the proportionately high unesterified sterol levels in molluscs is unknown.

E. PHARMACOLOGY AND ENDOCRINOLOGY

Thus far the endocrinology and pharmacology of molluscs appear to bear little resemblence to vertebrates with respect to hormonal control of carbohydrate or lipid metabolism (Martoja, 1972; Gilles, 1972). Attempts to alter molluscan blood sugar levels by insulin injection have given equivocal results (Goddard and Martin, 1966). Muscatine (1967) provided evidence of host chemical control of glycerol excretion by symbiotic algae in *Tridacna* (Bivalvia). Lubet and LeGall (1974) demonstrated the importance of cerebral ganglion and of neurosecretory cells on carbohydrate and lipid metabolism of the gastropod *Crepidula fornicata*. Following removal of the cerebroganglion or destruction of certain neuro-secretory cells, the carbohydrate and lipid levels fell. Replacement of the cerebral ganglion after 6 months resulted in partial recovery of lipid content but not of carbohydrate content which remained at the lower level. Additional control of nutrient

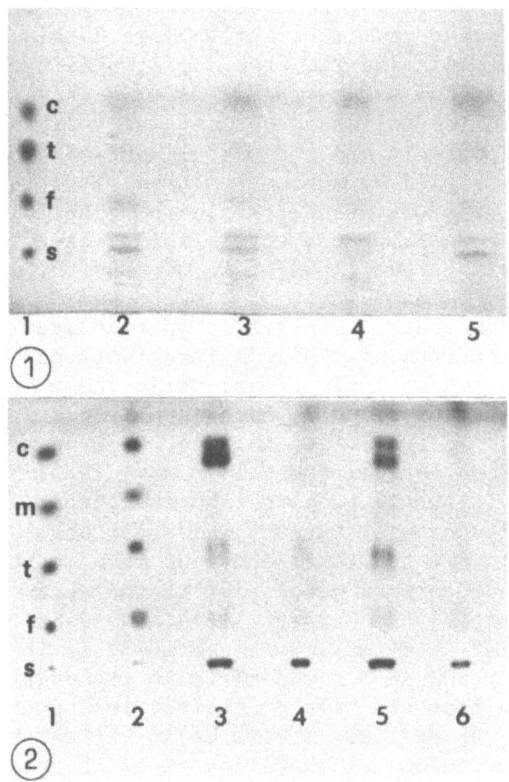

Fig. 1. Hemolymph lipids of *Nassarius obsoletus*. Silica gel thin-
        layer chromatogram developed in petroleum ether, diethyl
        ether, acetic acid (90/10/1), visualized with phospho-
        molybdic acid. C = cholesterol ester, t = triglyceride,
        f = free fatty acid, s = sterol. Lane 1 = standard,
        Lanes 2-5 = cell-free hemolymph lipids.

Fig. 2. Hemolymph lipids of *Mercenaria mercenaria*. Silica gel
        thin-layer chromatogram developed in petroleum ether, di-
        ethyl ether, acetic acid (85/15/1), visualized with
        molybdosilicic acid. C = cholesterol ester, m = methyl
        ester, t = triglyceride, f = free fatty acid, s = sterol.
        Lane 1 and 2 = standards. Lane 3 and 5 = blood cell
        lipids, Lanes 4 and 6 = cell-free hemolymph lipids.

distribution by the digestive gland was observed by Thompson
(1972a). The rate of nutrient transfer from the digestive gland
of *M. edulis* was independent of food concentration when food
was readily available. When food was not available, nutrient
release was governed by the concentration. An endocrinological
basis for regulation of carbohydrate and lipid levels remains
undemonstrated.

F. *NONNUTRIENT FUNCTION OF LIPIDS*

Lipids serve as structural components of membranes and as
nutrient stores. A respiratory role for carotenoids in molluscs
has recently been appreciated (Zs-Nagy, 1971; Zs-Nagy and
Ermini, 1972; Karnaukhov *et al.*, 1977). It has been suggested
that during anoxia the double bonds may serve as hydrogen
acceptors as does oxygen during aerobic electron transport
activity.

Lipid-rich larval molluscs may take advantage of the bouyancy
of their lipid stores during planktonic existence, although this
function has not been evaluated.

Specific adipose tissues do not occur in molluscs. Lipids as
fat deposits do not provide thermal insulation for invertebrates
as they do for vertebrates, especially mammals.

III. DIET AND LIPID METABOLISM

A. *FEEDING AND LIPID LEVELS*

Giese (1966, 1969) reviewed evidence showing amphinurans and
gastropods depend on lipid stores during periods of low food
availability whereas bivalves depend upon glycogen. Lawrence
(1976) discussed the role of lipids as stored nutrients in
molluscs and other invertebrates. Based on available evidence,
he concluded that lowered metabolic rates and general tissue
catabolism, rather than specific lipid catabolism, is often
the response of molluscs and other adult, benthic marine invertebrates to starvation. In temperate zones, decreased food
availability coincides with lowered temperature, therefore
lowered metabolic rates may be coincident with reduced food.
Certainly careful interpretation of relationships between
physiological condition, lipid levels and environmental parameters
is necessary. Lipid levels do vary with nutritional state as
expected if they are used as energy reserves even for some
species of bivalves and under appropriate conditions lipids have
nutritional as well as reproductive functions.

Digestive gland cells of bivalves accumulate lipid droplets
which are then depleted during starvation (Sumner, 1965). The
bivalves *Donax vittatus* and *Chlamys septemradiata* catabolize

stored lipid during starvation (Ansell, 1972, 1974). Gametogenesis and starvation reduce lipid levels in *Mytilus edulis* digestive gland (Thompson *et al.*, 1974). Biochemical determinations of tissue lipid levels confirm that lipid levels are reduced during periods of starvation in the absence of gametogenesis in *M. edulis*; however, over prolonged periods protein is the major energy source (Bayne, 1973b; Gabbott and Bayne, 1973). Lipid accounts for over half the caloric loss during autumn. During winter, when metabolic rate is increased due to gametogenesis, lipids are incorporated into eggs. Proteins then assume maximum importance in meeting somatic energy demands (Bayne, 1976).

When starved specimens of *M. edulis* were pulse-fed ^{14}C labelled algae, more label appeared in the eggs, mostly as lipid, than when well fed specimens were similarly given the labelled algae (Bayne *et al.*, 1975). Apparently in this species synthesized lipid is incorporated into eggs while protein from tissue resorption continues to sustain somatic metabolism-an example of intense devotion to successful reproduction.

Williams (1970) showed the gastropod *Littorina littorea* utilizes lipid during winter when less food is available. Barry and Munday (1959) concluded that *Patella vulgata*, (a gastropod), utilizes lipid primarily for gametogenesis and Blackmore (1969a, b) found no requirement for lipid by *P. vulgata* during winter. Van Brand *et al.* (1957) examined *Australorbis (=Biomphalaria) glabrata* after 30 days of starvation in water or 30 days of desiccation in air and found the lipid and carbohydrate levels were greatly reduced. Emerson (1967) demonstrated that a pulmonate, *Planorbis*, utilizes protein and lipid during starvation. After 58 days of starvation, 49% of the original protein and 22% of the lipid was consumed. Specimens of *Littorina planexis* utilized lipids almost exclusively during starvation (Emerson and Duerr, 1967). Obviously, a clear picture of the importance of lipid in meeting somatic energy demands of molluscs has not emerged.

Suryanarayanan and Alexander (1971, 1973) have shown that lipid is the major source for energy in the radular muscle of the pulmonate gastropod *Pila virens*. It is probable that lipid is a major energy source for muscular activity in molluscs as it is in vertebrates.

Elevated lipid levels in the digestive glands of bivalves reflect periods of active feeding (deMoreno *et al.*, 1976a; Ackman *et al.*, 1974) although few studies have been conducted under strictly controlled conditions. An exception is that of Oudejans and van der Horst (1974) who conclusively demonstrated that the pulmonate gastropod *Helix pomatia* increases deposited fat when placed on an oleic acid enriched lettuce diet compared with a nonenriched *ad libitum* lettuce diet. Helm *et al.* (1973) showed that dault diet influenced lipid content of the larvae of the oyster *Ostrea edulis*. Compared with well fed controls, starved oysters produced larvae with lower lipid content and reduced viability.

Lipid is unquestionably the major fuel reserve of larval molluscs. Millar and Scott (1967), Holland and Gabbott (1971), Gabbott and Holland (1973), Holland and Spencer (1973), Helm *et al.* (1973) and Bayne *et al.* (1975) have shown dramatic reduction of lipid levels of bivalve larvae during starvation. Holland *et al.* (1975) demonstrated that lipids are also the major nutrient stores of larval gastropods.

B. *FATTY ACIDS: MOLLUSCS AND VERTEBRATES COMPARED*

Fatty acid composition differs between mammals and molluscs. Marine molluscs, in particular, have high proportions of long chain polyunsaturated fatty acids. An interesting exception to this generality is the high concentration of $C_{22:6}$ synthesized from dietary $C_{18:3\omega3}$ (linoleic acid), in mammalian retinal phospholipids (Anderson, 1970). This fatty acid along with $C_{20:5}$ is characteristic of marine fatty acid profiles from diatoms to fish oils (Ackman, 1976; Malins and Wekell, 1969; Lee *et al.*, 1971). At least two fresh water invertebrates, the bivalve mollusc *Anodonta* and the crustacean *Orconectes*, contain these fatty acids in their phospholipids (Gardner and Riley, 1972; Wolfe *et al.*, 1965).

Unusual nonmethylene interrupted (NMID) fatty acids were discovered by Ackman and Hooper (1973) in lipids of the gastropod *Littorina littorea* and *Lunetta triseriata*, by Watanabe and Ackman (1972) in the European oyster *Ostrea edulis*, and by Pardis and Ackman (1975) in the American oyster *Crassostrea virginica*. These fatty acids are included in the peaks for methylene interrupted fatty acids of the same chain length during conventional gas-liquid chromatography. Pardis and Ackman (1975, 1977) stated that these could be of dietary origin or the results of noncarbon-alternating desaturation. Pardis and Ackman (1977) found NMID fatty acids in marine sea weeds. Aerobic bacteria are a major source of these fatty acids (Gurr and James, 1975); therefore, symbiotic organisms must also be considered as a possible source of NMID fatty acids found in molluscs. Watanabe and Ackman (1974) failed to find two of the NMID fatty acids earlier described from *O. edulis*.

C. *DIET AND MOLLUSCAN FATTY ACIDS*

Watanabe and Ackman (1972, 1974) examined the lipid content of specimens of *Crassostrea virginica* and *Ostrea edulis* after 6 hr of feeding on cultures of the algae *Isochrysis galbana* or *Dicrateria inornata*. The latter contains little $C_{20:5}$ and almost no $C_{22:6}$. The ratio of $C_{20:5}/C_{22:6}$ in total and polar lipids remained nearly the same in both oyster species as found in specimens on a natural diet. A slightly lower proportion of $C_{22:6}$

was observed in the triglyceride fraction of *C. virginica*. When
fed *Isochrysis galbana*, both oyster species increased the proportion of $C_{22:6}$ over that present in the alga. A conspicuous lack
of effect of dietary fatty acid was observed in both oyster
species with respect to the fatty acid composition of the total
lipids. The authors concluded that these two bivalve species
maintain species-specific ratios of $C_{22:5}/C_{22:6}$ and that *C.
virginica* increases its content of $C_{22:6}$ over dietary sources by
extention of $C_{20:5}$. Vertebrates also can interconvert $C_{20:5}$
and $C_{22:6}$ (Gurr and James, 1975).

When specimens of the bivalve *Arctica islandica* were held 10
weeks in flowing, filtered, natural sea water, they showed little
change in phospholipid levels of $C_{20:5}$ and $C_{20:6}$, whereas
sterol esters had slightly reduced levels of these two fatty
acids and triglycerides showed marked reduction (Ackman *et al.*,
1974). Almost no change was observed in digestive gland lipids
during these studies. Since these bivalves had similar fatty
acid patterns during active feeding in nature (May to July) and
during 10 weeks in filtered sea water, the authors concluded
that, like oysters, the fatty acid composition is species-specific.

The effect of starvation on tissue lipid fatty acid composition
was determined for *C. virginica* by Watanabe and Ackman (1977).
Specimens, stored at -2°C in air for up to 6 months, were
examined for quantitative and qualitative changes in total lipid
and fatty acid composition. From their data may be seen that
among major fatty acids the per cent composition of $C_{20:5}$ in the
phospholipid fraction increased during storage from 5.1% to 8.3%
and 6.2% to 12.9% in two experiments, while $C_{22:6}$ increased from
5.5% to 8.3% and from 7.4% to 15.4%. The relative proportion of
$C_{16:0}$, however, decreased slightly. Similar but small changes
of these fatty acids took place in the triglycerides. The authors
suggested that polyunsaturated fatty acids were conserved while
other fatty acids were preferentially catabolized. It should be
noted, however, that total lipids decreased at a rate slower than
the net dried weight loss. As a class, lipids were very little
used. The authors stated that under these conditions the specimens were presumed to be anaerobic. Utilization of fatty acids
under anaerobic conditions would seem unlikely since desaturation
is a necessary first step and desaturation reactions require
molecular oxygen. Thus, abandonment of lipid catabolism by the
bivalves under anaerobic conditions was possible.

The fatty acid composition of triglycerides of the marine
gastropod *Patella vulgata*, particularly $C_{22:4}$ and $C_{20:5}$, varied
with season which implies variations of diet (Gardner and Riley,
1972). Little change in phospholipid fatty acid composition
occurred during the same period.

Diet affected the amount of total lipid and the fatty acid composition of the marine bivalve *Mesodesma mactroides*. This clam contained less than 1% total lipid during periods of low food availability and over 2% lipid during summer algal blooms (deMoreno *et al.*, 1976a). The increased lipid was accounted for by increased nonpolar lipid in the digestive gland. The nonpolar lipid was rich in $C_{19:1}\omega9$, $C_{20:5}\omega3$, and $C_{22:6}\omega3$ fatty acids, which are characteristic of the algae. The concentration of $C_{22:6}$ in the total body fatty acids ranged from a maximum of 13% when algae were the major food to 4.3% when detritus was the major available food. Fluctuations among the phospholipids were not determined in this study.

M. mactroides absorbes linoleic and linolenic fatty acids directly from the sea water (deMoreno *et al.*, 1976a). During 6 hr exposure to ^{14}C linoleic or linolenic acids either dissolved directly in sea water or as incorporated into *Phaodactylium tricornutum*, the label appeared in the diglycerides but little or none appeared in the sterol ester, phospholipid, or triglyceride fractions. Labelled fatty acids extracted from clams fed algae previously exposed to $1^{14}C$-linolenic acid were in almost the same proportion to the labelled fatty acids in the algae. Between 50% and 75% of the label appeared unaltered by the clams. The ratio of $C_{20:5}\omega3/C_{22:6}\omega3$ was constant in the total lipids, fats, and phospholipids and equal to the ratio found in the diatoms. The authors concluded that in this species dietary fatty acids regulated fatty acid composition. The authors interpreted their findings as indirect evidence of lipid synthesis via glycerophosphate through phosphatidic acid to diglycerol and other lipids. The availability of fatty acids is known to regulate the composition of rat adipose triglycerides which are principally synthesized via the $\omega f/\propto$ glycerophosphate pathway (Christie *et al.*, 1974). Conversion of the diglycerides to other lipids by specimens of *M. mactrodes* is considerably slower than diglyceride formation from ingested fatty acids. Also, since fatty acid conversion included elongation of linoleic acid to $C_{22:2}\omega6$ but not to linoleic or arachandonic acid, deMoreno *et al.* (1976b), suggested that $\Delta6$ desaturation did not occur in this species.

D. *LIPID REQUIREMENTS: MOLLUSCS AND VERTEBRATES COMPARED*

In certain respects molluscan lipid metabolism appears more similar to that of mammals than to that of insects, the invertebrate group for which the most information is available. Fatty acid composition of insects varies with diet within relatively narrow species-characteristic distribution except that polyunsaturated fatty acids are present in adult insects only when present in the diet (Thompson, 1973). Both molluscs and mammals are capable of at least some *de novo* synthesis of polyunsaturated fatty acids.

Fatty acid metabolism in vertebrates is directly dependent upon diet as a source of linoleic acid ($C_{18:2}\omega_6$) and to a lesser extent, linolenic acid ($C_{18:3}\omega_6$) and aracadonic acid ($C_{20:4}\omega_6$). This need arises because vertebrates lack the enzymes to perform desaturation at the $\Delta 12$-13 position of oleic acid and, in general, can only desaturate between an existing double bond and the carboxyl end of the fatty acid (Gurr and James, 1975). Only linoleic acid cannot be synthesized at all: linolenic and aracadonic acids can be synthesized from linoleic acid but at rates insufficient to meet metabolic demands.

Normal rat brain development, including phospholipid synthesis, is dependent upon dietary essential fatty acids (Rathbone, 1965; Houtsmueller, 1972; Odutuga, 1977). All EFA deficiency symptoms in rats can be alleviated by feeding aracadonic acid, but linoleic and linolenic acid levels remain low (Rahm and Holman, 1964). In the absence of all other dietary lipids except lino-lenic acid, rat retina become depleted of $C_{22:6}$ (Tinoco *et al.*, 1977). The absence of linolenic acid from the diet in the presence of linoleic acid results in no evidence of EFA deficiency other than the reduction of retinal $C_{22:6}$. EFA requirements of vertebrates are related to metabolism of polyenoic fatty acids characteristic of marine molluscs.

Cod liver oil, rich in linolenic acid, promoted better growth in specimens of *C. virginica* than did corn oil which is low in this fatty acid (Castell and Trider, 1974). There may be an essential fatty acid requirement in this species, but additional evidence is needed.

Van der Horst (1973) found that the pulmonate *Cepaea nemoralis* can synthesize linoleic acid ($C_{18:2}\omega_6$) from acetate. This species has a very complete fatty acid synthetic and regulatory ability. In fact, most dietary linolenic and linoleic acids are metabolized to CO_2 (van der Horst and Oudejans, 1976). Although some fatty acids may be incorporated directly into triglyceride, they are rapidly altered to species-specific fatty acids when incorporated into phospholipids. The snail tends to conserve stores of fatty acids usually considered essential but it acquires them primarily by *de novo* synthesis and does not directly incorporate significant amounts of even major structural fatty acids directly into its lipids. The primary product of *de novo* synthesis is palmitate ($C_{16:0}$), which is elongated and desaturated to oleic acid ($C_{18:1}$) and linoleic acid ($C_{18:2}$). Further desaturation and elongation to $C_{20:2}$, $C_{20:3}$, and $C_{20:4}$ are also accomplished by specimens of *C. nemoralis*. Synthesis of fatty acids occurs at reduced rates during anoxia (van der Horst, 1974) and hibernation (van der Horst *et al.*, 1974). This mollusc, like vertebrates, readily synthesizes

palmitate *de novo*. It evidently excels vertebrates by possessing
a complete fatty acid synthesizing system with which it maintains
its fatty acid composition virtually autonomous from its diet.
The earlier observations by van der Horst and Zandee (1973) that
C. nemoralis specimens maintain constant fatty acid composition
throughout the year may be explained by the extraordinary fatty
acid synthesizing ability of this species.

Essential fatty acids or their derivatives are utilized for
vertebrate prostaglandin synthesis. The occurrence of the
prostaglandin PGB_2 in the gastropod *Cyphoma gibbosum* has been
reported (Steudler *et al.*, 1977). The authors hypothesize that
prostaglandin PGA_2 ingested from gorgonian corals serves as the
precursor.

Clinical studies of prostaglandins in humans have shown rapid
uptake of PGE_2 and PGF_2 following oral administration (Karim,
1971a,b). Negative feedback of biosynthesis has been discussed
(Lands *et al.*, 1974). Dietary essential fatty acids have been
shown to regulate the rate of prostaglandin synthesis (Hemler
and Lands, 1977). There seems to be no information concerning
the relationship of prostaglandins normally ingested with the
diet and prostaglandin levels in vertebrates. In view of the
example provided by *C. gibbosum*, such studies seem warranted.

Sphingolipids contain sphingosine rather than glycerol as the
alcohol. Two pathways for synthesis, acylation of sphingo-
phosphoryl choline and addition of the phosphate group to acyl
spingosine, are commonly recognized, although their relative
importance *in vivo* is not known (Gurr and James, 1975). It is
generally accepted that the long chain bases are synthesized
in vivo (Rossiter and Strickland, 1960).

The fatty acids and bases of the adductor muscle of the oyster
Crassostrea gigas contain only straight chain bases and 94.3%
hexadecanoic (C_{16}) and 5.4% heptadecanoic (C_{17}) fatty acids
(Matsubara and Hayashi, 1973). Specimens of the freshwater
bivalve *Corbicula sandai* contain branched as well as straight
chain bases (Sugita *et al.*, 1976). Thus at least this mollusc,
like mammals, apparently can synthesize both the straight and
branched chain bases. Further analogies between these two
groups with respect to sphingolipid metabolism await additional
study.

E.   FATTY ACID REGULATION: MOLLUSCS AND VERTEBRATES COMPARED

The fatty acid composition of structural lipids (phospholipids)
is of critical importance in establishing cell membrane properties
such as permiability. The mechanisms by which fatty acid compositions of various phospholipid classes are regulated in mammals
are not clearly understood, although dietary content is obviously important. Selection of diglycerides containing specified
fatty acids followed by synthesis of the specific phospholipid

or exchange of fatty acids by trans-acylases subsequent to
phospholipid synthesis have both been demonstrated (Gurr and
James, 1975).

Fatty acid composition of phospholipids of ectothermic
organisms may be related to environmental temperature (Malins
and Wekell, 1969). However, van der Horst *et al.* (1974),
working with the pulmonate *Cepaea nemoralis*, and de Moreno *et al.*
(1976b), working with field collected specimens of the bivalve
Mesodesma mactroides, found that temperature did not affect
phospholipid fatty acid composition.

Oudejans and van der Horst (1974) reported highly specific
phospholipid fatty acid composition of the pulmonate *Helix
pomatia*. The deposition of lipids of this snail was influenced
by the fatty acid composition of the diet, but fatty acids were
not present in proportion to those in the diet. The major fatty
acid in the diet, $C_{18:3\omega3}$, was in low concentration in the depot
lipids. Van der Horst *et al.* (1973) injected $1^{14}C$-acetate into
Cepaea nemoralis. Phospholipids were highly labelled with species-
specific fatty acids. The acetate was incorporated into $C_{16:0}$
and $C_{18:0}$ which apparently served as the major substrates from
which the specific fatty acids were formed. Pyruvate, glucose
and alanine, as well as acetate, were rapidly assimilated into
the pool of saturated fatty acids. During hibernation, injected
acetate was also incorporated into $C_{16:0}$ and $C_{18:0}$ followed by
aerobic synthesis of species-specific fatty acids in the phospholipid fraction. The acetate was incorporated into $C_{18:0}$ most
readily in hibernating snails (van der Horst *et al.*, 1974),
although $C_{16:0}$ was reported as the most heavily labelled saturated
fatty acid in the nonhibernating snails (van der Horst, 1973).
The only other observed difference in synthetic activity of awake
and hibernating snails was the greatly reduced rate in the
latter. The authors calculated that 1 week of synthetic activity
of hibernating snails was equivalent to ½ hr in nonhibernating
specimens.

When the role of dietary lipids in molluscs is compared with
vertebrates, it appears from the foregoing discussion that
certain molluscs may be the better regulators of structural
lipids. Gudbjarnason and Oskarsdottier (1977) have shown that
fatty acid composition of rat cardiac phospholipid is profoundly
affected by dietary fatty acids. When rats were placed on a
diet rich in cod liver oil, the neutral lipids showed moderate
increase in the characteristic oils but the phospholipids showed
up to 30% replacement of n-6 fatty acids by the marine n-3 fatty
acids. Specifically, 22:6 replaced 18:2, 18:0, and 20:4 in
these structural lipids. In another study, tissue phospholipids
reflected dietary fatty acids though in a less pronounced manner
than did triglycerides (Christie *et al.*, 1974). Phospholipids of

mouse hepatocyte plasma membranes also reflected dietary fatty acid composition (Hopkins and West, 1977).

Ingestion of rapeseed oil, rich in $C_{22:1}\omega f/9$, leads to fatty deposits and necrotic lesions of the myocardium (Beare-Rogers and Nera, 1972). This fatty acid also inhibits cardiac mito-chondrial respiration *in vitro* (Hsu and Kummerow, 1977). Christiansen *et al.* (1977) have discussed the inhibitory effects of hydrogenated (monoethylenic) marine fatty acids on vertebrate fatty acid oxidation *in vitro*. Certain fatty acids can also inhibit vertebrate cholesterol syntheses *in vitro* (Kuroda and Endo, 1977; Faas *et al.*, 1977).

With respect to the incorporation and metabolism of fatty acids, molluscs provide interesting examples of regulation of phospholipid composition and of toleration of high levels of polyenoic C_{20} and C_{22} fatty acids. In fact, these fatty acids appear to be necessary to molluscs since they are present in fresh water as well as marine species (Gardner and Riley, 1972).

Requirements for structural phospholipid composition differ between molluscs and mammals since mammals apparently regulate palmitate and oleate (Keenan and Morre, 1970), but are not as able to regulate polyenoic fatty acids. Cardiac phospholipids in particular, exhibit a profound change in fatty acid composi-tion when the diet is rich in $C_{21:5}$ and $C_{22:6}$. The observations that molluscs can regulate the levels of $C_{20:5}$ and $C_{22:6}$ are of particular interest with respect to the dependency upon dietary linoleic acid for normal rat retinal lipid composition. The effects of varying dietary $C_{18:3}$, $C_{20:5}$, and $C_{22:6}$ should be compared between these groups.

F. ETHER LIPIDS: MOLLUSCS AND VERTEBRATES COMPARED

The steps in the synthesis of glycerol ethers (ether containing triglycerides) and plasmologens (ether containing phospholipids) are fairly well understood (Synder, 1972). Plasmologen synthe-sis involves desaturation of the fatty acids in ether linkage with phospholipid. Intestinal epithelium cells, brain cells, and ascites tumor cells contain the enzymes required to perform this desaturation (Blank *et al.*, 1970). Since the first report of plasmologens in molluscs by Theile (1959), these organisms have been utilized as models to study the steps involved in the synthetic pathways of ether lipids (Thompson, 1966, 1968, 1972a). However, the relationship between the ether lipid content and environmental parameters including food is not clear. Theile (1959) described seasonal variations in levels of scallop plasmologens. Rapport (1961) and Rapport and Alonzo (1960) analyzed levels of plasmologens in species of cephalopods, bivalves, and gastropods, and found higher levels than occur in mammalian brain. Isay *et al.* (1976) have analyzed the glyceryl

ether content, as per cent total lipid, from 22 molluscan species
occurring in the Sea of Japan and eight tropical species. The
variation in content among species within each class was high
so that neither geographic or taxonomic relationships were
established. The composition among 16 Japan Sea bivalves
ranged from 1.4% to 9.9% while the two tropical species contained 1.5% to 16.2% glyceryl ether. Although control mechanisms
regulating the alkyl glyceryl ether levels have not been found,
there is a negative feedback between ingestion of these lipids
and *de novo* synthesis from ^{14}C palmitate (Thompson, 1972a).
Thompson (1972b) suggests that these compounds aid in stabilizing
membrane surfaces since they are relatively insensitive to
lipases. Thus, although the biochemical pathways of synthesis
of certain ether lipids have been elucidated, partly by use of
molluscan models, there exist no specific data relating ether
lipids in diet and synthetic activity in mammals and only one
study of the effect of diet on synthesis among the molluscs.

Since there are increased levels of glyceryl ethers in vertebrate neoplasms (Synder, 1969a,b; Albert and Anderson, 1977),
further investigations of metabolism of these lipids, particularly of a comparative nature between molluscan species with high
and low levels, may increase an understanding of vertebrate
tumor metabolism.

G. DIET AND MOLLUSCAN STEROLS

The major sterol of higher animals is cholesterol and the
complete pathway of cholesterol synthesis from acetate is known.
The ability to synthesize cholesterol *de novo* is probably universal among vertebrates, although this capability may be
lacking in other animal groups. Insects are unable to synthesize
sterols and depend upon dietary sources, or, as in the case of
Xyleborus ferrugineus, upon symbiotes. Ergosterol, required for
successful reproduction of this beetle, is supplied by ecto-
symbiotic fungi (Kok and Norris, 1973).

It is well known that vertebrates synthesize cholesterol in
inverse proportion to dietary intake. Other effects of diet
include increase in serum and hepatic cholesterol among vitamin
C-deprived guinea pigs (Ginter *et al.*, 1973) which, like man,
require exogenous sources of this vitamin.

Sterol analysis of molluscs has received continuing attention
(Voogt, 1972; Idler and Wiseman, 1971, 1972). Recent work
confirms previous reports of the variety in sterols, at least
among marine molluscs (Khalil and Idler, 1975; Patterson *et al.*,
1975; Piretti and Viviani, 1976), which contrasts with the
predominance of the single sterol, cholesterol, among vertebrates.

Individual species of molluscs contain small amounts of
many sterols which differ in location and number of double bonds
and in the structure of the side chain. Modification of the

sterol ester side chain by desaturation has been demonstrated in the bivalve *Saxidomus giganteus* by Fagerlund and Idler (1961).

The absolute amount of sterols as per cent of tissue weight is higher than in mammalian muscle. Ackman (1976) reports, for example, cholesterol values of 35-50 mg/100 gm tissue and Watanabe and Ackman (1977) report levels of 145 mg total sterol/ 100 gm whole tissue, although Voogt (1975a,b) found crude sterol values in bivalves ranging between 5 mg/100 gm (in *Atrina fragilis*) to 150 mg/100 gm (in *Cardium edule*). These values can be compared with 65-58 mg/100 gm in beef (Feeley *et al.*, 1972). It is generally accepted that sterols are important structural components of cell membranes. Sterols are precursors for vertebrate steroid hormones. Steroid hormones also occur in pulmonates (Gottfried and Dorfman, 1970) and bivalves (Hagerman *et al.*, 1957; Idler *et al.*, 1969; Saliot and Barbier, 1971). Steroid hormone synthesis has been examined in *Mytilus edulis* gonad by de Longcamp *et al.* (1974). These authors found incorporation of ^{3}H pregnenolone into progesterone, although they did not detect labelling of cholesterol or 7-dehydrocholesterol from ^{3}H-sodium acetate nor did they observe incorporation of ^{3}H-acetate or 4-^{14}C cholesterol into steroid hormones. The tritium label was lost from the cholesterol fraction during repurification and was not detected in the hydrocholesterol fraction. Apparently this molluscan species is capable only of higher steps in synthesis of gonadal hormones and must require a dietary source of pregnenolone or other modified cholesterol nucleus.

Conflicting results have been obtained concerning the sterol synthesizing abilities of bivalves. Thus Voogt (1972) reported that 1^{14}C acetate incorporation into the sterol fraction was not detected for the bivalves *Ostrea edulis, Mya arenaria*, or *Cyrina islandica*. However, following repeated injection of 1^{14}C acetate for 7 days, some 3β sterols of *Ostrea edulis* were labelled, proving that a low level of synthetic activity occurs. Salaque *et al.* (1966) found that neither 2^{14}C mevalonate or methyl ^{14}C-methionine were incorporated into sterols of the oyster *Ostrea gryphea*. Voogt (1972), on the other hand, reported that specimens of *Cardium edule* utilized 2^{14}C-acetate but not 1^{14}C-acetate for sterol synthesis. Fagerlund and Idler (1960) observed sterol synthesis from 2^{14}C-acetate in *Mytilus californicus* and *Saxidomus giganteus*. Voogt (1975b), again, found incorporation of 2^{14}C-acetate but not 1^{14}C-acetate into sterols of several bivalves which was probably the result of alkylation rather than *de novo* synthesis.

De Longcamp *et al.* (1974) concluded that bivalve sterols are of dietary origin and are modified by the organism but Tamura *et al.* (1974) concluded that sterol levels of *Crassostrea virginica* remained constant on a sterol-free diet. This could only happen if sterols are conserved to a remarkable degree by these oysters or if they are able to synthesize sterols.

Table 2. Sterol Synthesis of Some Representative Molluscs

Mollusc	Precursor	Sterol	Author
Amphinura			
Liolophura japonica	mevalonate	7-cholesterol, cholesterol	Teshima & Kanazawa (1973)
Gastropoda			
Natica cataena	acetate	no sterols	Voogt (1971)
Natica cataena	mevalonate	sterols	Voogt (1971)
Crepidula formacata	acetate	squalene & sterols	Voogt (1971)
Crepidula formacata	mevalonate	squalene & sterols	Walton & Pennock (1972)
Littorena littorea	mevalonate	squalene & sterols	Walton & Pennock (1972)
Gibbula cineraria	mevalonate	squalene & sterols	Walton & Pennock (1972)
Nucella lapilus			
Planorbarius corneus,			
Patella vulgata,			
Viviparus fisciatus,			
Littorina littorea	acetate	sterols	Voogt (1968a,b, 1969)
Buccinium undatum	acetate	no sterols	Voogt (1972)
Buccinium undatum	acetate, mevalonate	squalene & sterols	Voogt (1972)
Buccinium undatum	lanosterol	cholesterol	Khalil & Idler (1976)
Littorina littorea	demosterol	cholesterol	Idler & Wiseman (1971)
Helix pomatia	acetate	cholesterol	Addink & Ververgaert (1963)
Haliotis gurneri	mevalonate	cholesterol	Teshima & Kanazawa (1974)
Ariolimac californicus	acetate	cholesterol	Gottfried & Dorfman (1969)
Arion rufus	acetate	3β sterol	Voogt (1967a)
Aplysia depilans	acetate	cholesterol	dePrisco *et al.* (1973)
Cepaea nemoralis	acetate	sterol	van der Horst *et al.*(1974)

Table 2. (Continued)

Mollusc	Precursor	Sterol	Author
Bivalvia			
Mytilus californicus	acetate	Δ^5 3β sterol	Fagerlund & Idler (1960)
Saxidomus giganteus	acetate	Δ^5 3β sterol	Fagerlund & Idler (1960)
Mytilus edulis	acetate	no cholesterol	deLongcamp *et al.* (1974)
Mytilus edulis	mevalonate	no sterols	Walton & Pennock (1972)
Cardium edule	mevalonate	no sterols	Walton & Pennock (1972)
Ostrea graphea	mevalonate	no squalene or sterol	Salaque *et al.* (1966)
Mytilus edulis	acetate	no squalene or sterol	Voogt (1975a)
Atrina fragilis	acetate	no squalene or sterol	Voogt (1975a)
Ostrea edulis	acetate	no squalene or sterol	Voogt (1975a)
Anodonta cygnea	acetate 1^{14}C:2^{14}C	no sterols: sterols	Voogt (1975b)
Cardium edule	acetate 1^{14}C:2^{14}C	no sterols: sterols	Voogt (1975b)
Cyprina islandica	acetate 1^{14}C:2^{14}C	no sterols: sterols	Voogt (1975b)
	mevalonate	no sterols	
Mya arenaria	acetate 1^{14}C:2^{14}C	no sterols: sterols	Voogt (1975b)
Cardium edule	acetate 1^{14}C:2^{14}C	no sterols: sterols	Voogt (1972)

Voogt (1975a,b) has examined sterol synthesis in several bivalve species and found species differ in capability (Table 2). Walton and Pennock (1972) found that *Mytilus edulis* and *Cardium edule* were unable to form squalene or sterol from 2^{14}C-mevalonate, although the gastropods *Nucella lapillus*, *Litterina littorea*, and *Gibbula cineraria* could. However, Teshima and Kanazawa (1974) reported incorporation of 2^{14}C-mevalonate into cholesterol, 22 dehydrocholesterol, demonsterol, and 24 demosterol by *Mytilus edulis* and into cholesterol by the gastropod *Haliotis gurneri*.

Khalil and Idler (1976) observed biosynthesis of cholesterol from ^{3}H-lanosterol in the marine gastropod *Buccinium undatum*. Voogt (1967b) had reported the species unable to synthesize 3β sterol from 1^{14}C-acetate, although he observed incorporation of labelled carbon from 2^{14}C-acetate.

Zandee (1967) reported a cephalopod, *Sepia officinalis*, is unable to synthesize any sterols.

Teshima and Kanazawa (1973) found that the polyplacophoran *Liolophura japonica* contains only 3% cholesterol and 97% 7-cholestanol. This species incorporated 2^{14}C-mevalonate almost exclusively into 7-cholestanol. These authors suggested a cholesterol biosynthetic pathway involving interconversion of 7-cholestanol to cholesterol as reported from echinoderms and rats.

DiPrisco *et al.* (1973) demonstrated that the opisthobranch *Aplysia depilans* can synthesize cholesterol from 1^{14}C-acetate. Gottfried and Dorfman (1969) reported *de novo* sterol synthesis in two terrestrial pulmonate species.

Kanazawa *et al.* (1976) compared sterols from coral reef molluscs to noncoral reef species. Their study provided indirect evidence for dietary origins of sterols. Coral reef invertebrates including one gastropod and two bivalve species had sterol compositions which included more 28 carbon sterols than noncoral reef species in which cholesterol was the major sterol and 24 methylene cholesterol was second in amount. These authors reasoned that since the sterol compositions of coral reef invertebrates at different trophic levels were similar, these sterols must be passed up the food chain with only minor alterations. Similar conclusions were reached by Steudler *et al.* (1977).

A dietary requirement of molluscs for sterols, based upon lack of ability to synthesize sterol, has not been clearly established.

The artificial diets employed by Castell and Trider (1974) apparently, lacked sterol. Control animals, feeding on natural plankton, which contain sterol (Tornabene *et al.*, 1974; Goad *et al.*, 1972), grew better than any of the experimental specimens.

The question of the role of symbiotic organisms in supplying sterols to molluscs has not been addressed and studies directed towards regulation of sterol synthesis over prolonged periods by varying diet are needed.

CONCLUSIONS

Detailed knowledge of the lipid metabolism of even one mollus-can species is not available except that the pulmonate *Cepaea nemoralis* is apparently capable of complete *de novo* fatty acid synthesis. The fragmentary data on gross alterations in lipid content or level related to various factors including starvation are generally from natural or at best relatively uncontrolled conditions. For example, aquatic specimens often have been held in filtered natural pond or sea water, making unequivocable interpretation impossible.

The improvement in growth of bivalves when diets were supplemented with corn oil and cod liver oil (Castell and Trider, 1974) indicates the importance of considering the lipid content of the diet.

It is imperative to establish whether any species other than *Cepaea nemoralis* is capable of *de novo* fatty acid synthesis, including synthesis of fatty acids regarded as essential in the vertebrate diet. The intrinsic and extrinsic factors governing cholesterol metabolism must be established. Regulation of synthetic ability, rates of synthesis, and interconversions must be considered with attention to the possible role of symbiotes and of body surface absorption from natural water. Only laboratory reared or other specimens known to be free of parasites should be used for such studies and consideration of protozoan and prokaryotic symbiotes, as stated above, is warranted.

Except for the studies by Oudejans and van der Horst (1974), reports of alterations or lack of alterations of phospholipid fatty acid composition of molluscs have been based on short-term (6 hr) dietary changes or in the wild where the reduced availability of a particular fatty acid was concurrent with relative increases in other fatty acids.

Studies on mammals have shown dietary influences on fatty acid composition with some tissues being more affected than others (Carroll, 1965).

Insufficient data are available on molluscs to reach firm conclusions concerning differences in fatty acid metabolism of different tissues and findings of stability or variability of phospholipid composition must be evaluated with the experimental conditions in mind. It does seem reasonable, however, to conclude that in both mammals and molluscs depot lipids are more affected by diet than are structural lipids.

Information of direct relevancy to vertebrate systems has come from studies of ether lipid metabolism and additional studies should be of significance to a general understanding of the formation and regulation of these lipids.

The synthesis and regulation of other lipids vary among species and examples must be chosen with care. In many instances the synthetic ability will have to be first established before studies on diet regulation can be made.

ACKNOWLEDGMENT

The author thanks Nora Hartley for her patience in typing the draft and final copy of this manuscript.

REFERENCES

Ackman, R. G. (1976). Fish oil composition. Objective Methods for Food Evaluation. *Nat. Acad. Sci.*, 103-131.

Ackman, R. G., Epstein, S., and Kelleher, M. (1974). A comparison of lipids and fatty acids of the ocean quahaug, *Arctica islandica*, from Nova Scotia and New Brunswick. *J. Fish. Res. Bd. Can.*, 31, 1803-1811.

Ackman, R. G. and Hooper, S. N. (1973). Non-methylene-interrupted fatty acids in lipids of shallow-water marine invertebrates: a comparison of two molluscs (*Littorina littorea* and *Lunatia triseriata*) with the sand shrimp (*Cragon septemspinosis*). *Comp. Biochem. Physiol.*, 46B, 153-165.

Addink, A. D. F. and Ververgaert, P.H.J.T. (1963). Biosynthesis of cholesterol and fatty acids in a snail *Helix pomatia* L. after administration of $1-{}^{14}C$-acetate. *Archs. int. Physiol. Biochem.*, 71, 797-801.

Albert, D. H. and Anderson, C. E. (1977). Ether-linked glycerolipids in human brain tumors. *Lipids* 12, 188-192.

Allen, W. V. (1977). Interorgan transport of lipids in the blood of the gumboot chitin *Cryptochiton stelleri* (Middendorf). *Comp. Biochem. Physiol.*, 57A, 41-46.

Anderson, R. E. (1970). Lipids of occular tissues, IV. A comparison of the phospholipids from the retina of six mammalian species. *Exp. Eye Res.*, 10, 339-344.

Ansell, A. D. (1972). Distribution, growth and seasonal changes in biochemical composition for the bivalve *Donax vittatus* (DaCosta) from Kames Bay, Millport. *J. Exp. Mar. Biol. Ecol.*, 10, 137-150.

Ansell, A. D. (1974). Seasonal changes in biochemical composition of the bivalve *Chlamys septemradiata* from the Clyde Sea area. *Mar. Biol.*, 25, 85-99.

Barry, R. J. C. and Munday, K. A. (1959). Carbohydrate levels in *Patella*. *J. Mar. Biol. Ass. U.K.*, 38, 81-95.

Bayne, B. L. (1973a). Physiological changes induced in *Mytilus edulis* L. induced by temperature and nutritive stress. *J. Mar. Biol. Ass. U.K.*, 53, 39-58.

Bayne, B. L. (1973b). Aspects of the metabolism of *Mytilus edulis* L. during starvation. *Neth. J. Sea Res.*, 7, 399-410.

Bayne, B. L. (1976). Marine Mussels: their Ecology and Physiology. *Cambridge University Press*, Cambridge, 506 pp.

Bayne, B. L., Gabbott, P. A. and Widdows, J. (1975). Some effects of stress in the adult on the eggs and larvae of *Mytilus edulis* L. *J. Mar. Biol. Ass. U.K.*, 55, 675-689.

Beare-Rogers, J. L. and Nera, E. A. (1972). Cardiac fatty acids and histopathology of rats, pigs, monkeys and gerbils fed rapeseed oil. *Comp. Biochem. Physiol.*, 41B, 793-800.

Blackmore, D. T. (1969a). Studies on *Patella vulgata* L. I. Growth reproduction and zonal distribution. *J. Exp. Mar. Biol. Ecol.*, 3, 200-213.

Blackmore, D. T. (1969b). Studies on *Patella vulgata* L. II. Seasonal variation in biochemical composition. *J. Exp. Mar. Biol. Ecol.*, 3, 231-245.

Blank, M. L., Wykle, R. L., Piantadosi, C., and Synder, F. (1970). The biosynthesis of plasmologens from labeled 0-aklyl-glycerols in Ehrlich Ascites cells. *Biochim. Biophysics. Acta*, 210, 442-447.

Carroll, K. K. (1965). Dietary fat and fatty acid composition of tissue lipids. *J. Am. Oil Chem. Soc.*, 42, 516-552.

Castell, J. D. and Trider, D. J. (1974). Preliminary feeding trials using artificial diets to study the nutritional requirements of oysters *(Crassostrea virginica)*. *J. Fish. Res. Board Can.*, 31, 95-99.

Cheng, T. C. (1965). Histochemical observations on changes in the lipid composition of the American oyster, *Crassostrea virginica* (Gmelin), parasitized by the trematode *Bucephalus* sp. *J. Invert. Pathol.*, 7, 398-407.

Cheng, T. C. (1967). Marine molluscs as hosts for symbioses: With a review of known parasites of commercially important species. *Advan. Mar. Biol.*, 5, 1-424.

Cheng, T. C. and Lee, F. O. (1971). Glucose levels in the mollusc *Biomphalaria glabrata* infected with *Schistosoma mansoni*. *J. Invertebr. Pathol.*, 18, 395-399.

Christiansen, R. Z., Christopherson, B. O., and Bremer, J. (1977). Monoethylenic C_{20} and C_{22} fatty acids in marine oil and rapeseed oil. Studies on their oxidation and on their relative ability to inhibit palmitate oxidation in heart and liver mitochondria. *Biochim. Biophys. Acta*, 487, 28-36.

Christie, W. W., Moore, J. H., and Gottenbos, J. J. (1974). Effect of dietary saturated fatty acids and linoleic acid upon the structures of triglycerides in rabbit tissues. *Lipids*, 9, 201-207.

deLongcamp, D., Lubet, P., and Drosdowsky, M. (1974). The *in vitro* biosynthesis of steroids by the gonad of the mussel *(Mytilus edulis)*. *Genl. and Comp. Endocrin.*, 22, 116-127.

deMoreno, J. E. A., Moreno, V. J., and Brenner, R. R. (1976a).
    Lipid metabolism of the yellow clam, *Mesodesma mactroides:*
    1. Composition of the lipids. *Lipids,* 11, 334-340.
deMoreno, J. E. A., Moreno, V. J., and Brenner, R. R. (1976b).
    Lipid metabolism of the yellow clam, *Mesodesma mactroides:*
    2 - polyunsaturated fatty acid metabolism. *Lipids,* 11, 561-
    566.
diPrisco, C. L., Fulgheri, F. D. and Tomasucci, M. (1973).
    Identification and biosynthesis of steroids in the marine
    mollusc *Aplysia depilans*. *Comp. Biochem. Physiol.,* 45B, 303-
    310.
Emerson, D. N. (1967). Carbohydrate oriented metabolism of
    Planorbis corneus (Mollusca, Planorbidae) during starvation.
    Comp. Biochem. Physiol., 22, 571-579.
Emerson, D. N. and Duerr, F. G. (1967). Some physiological
    effects of starvation in the intertidal prosobranch *Littorina
    planexis* (Philippi, 1847). *Comp. Biochem. Physiol.,* 20, 45-53.
Epifanio, C. E. (1976). Culture of Bivalve Mollusks in Re-
    circulating systems: Nutritional Requirements. *In:* "*Proc.
    First Int. Conf. on Aquaculute Nutrition*" pp. 173-194.
    (K.S. Price, Jr., W. N. Shaw, K. D. Danberg, eds.).
    University of Delaware, Newark, Delaware.
Faas, E. H., Carter, W. J., and Wynn, J. O. (1977). Unsaturated
    fatty acyl-CoA inhibition of cholesterol synthesis *in vitro*.
    Biochim. Biophys. Acta., 487, 277-286.
Fagerlund, U. H. M. and Idler, D. R. (1960). Marine sterols.
    Sterol biosynthesis in molluscs and echinoderms. *Can. J.
    Biochem. Physiol.,* 38, 997-1002.
Fagerlund, U. H. M. and Idler, D. R. (1961). Biosynthesis of 24-
    methylene cholesterol in clams. *Can. J. Biochem. Physiol.,*
    39, 1347-1355.
Feeley, R. M., Criner, P. E., and Watt, B. K. (1972). Cholesterol
    content of foods. *J. Am. Diet Assoc.,* 61, 134-149.
Gabbott, P. A. and Holland, D. L. (1973). Growth and metabolism
    of *Ostrea edulis* larvae. *Nature,* 241, 475-576.
Gabbott, P. A. and Bayne, B. L. (1973). Biochemical effects of
    temperature and nutritive stress on *Mytilus edulis* L.
    J. Mar. Biol. Ass. U.K., 53, 269-286.
Gardner, D. and Riley, J. P. (1972). The component fatty acids
    of the lipids of some species of marine and fresh water
    molluscs. *J. Mar. Biol. Ass. U.K.,* 52, 827-838.
Giese, A. C. (1966). Lipids in the economy of marine inverte-
    brates. *Physiol. Rev.,* 46, 244-298.
Giese, A. C. (1969). A new approach to the Biochemical Composition
    of the mollusc body. *In:* "*Oceanogr. Mar. Biol. Ann. Rev.*"
    p. 175-229. (H. Barnes, ed.). George Allen & Unwin, Ltd.,
    London.

Gilles, R. (1972). Biochemical ecology of mollusca. *Chem. Zool.*, 7, 467-499.

Ginter, E., Nemec, R., Cerven, J., and Milkus, L. (1973). Quantification of lowered cholesterol oxidation in guinea pigs with latent vitamin C deficiency. *Lipids*, 8, 135-141.

Goad, L. J., Knapp, F. F., Lenton, J. R., and Goodwin, T. W. (1972). Sterol side-chain alkylation mechanism in a *Trebouxia* species. *Biochem. J.*, 129, 219-222.

Goddard, C. K. and Martin, A. W. (1966). Carbohydrate metabolism. *In:* "Physiology of the Mollusca." Vol. 2. pp. 275-308. (K. M. Wilber and C. M. Yonge, eds.). Academic Press, New York.

Goodman, D. S. (1965). Cholesterol ester metabolism. *Physiol. Rev.*, 45, 747-839.

Gottfried, H. and Dorfman, R. I. (1969). The occurrence of the *in vivo* cholesterol biosynthesis in an invertebrate *Ariolimax californicus*. *Gen. Comp. Endocrinol.*, (Suppl.) 2, 590-593.

Gottfried, H. and Dorfman, R. I. (1970). Steroids of invertebrates. V. The *in vitro* biosynthesis of steroids by the male-phase ovotestis of the slug *(Ariolimax californicus)*. *Gen. Comp. Endocrinol.*, 15, 120-138.

Gudbjarnason, S. and Oskarsdottier, G. (1977). Modification of fatty acid composition of rat heart lipids by feeding cod liver oil. *Biochim. Biophys. Acta.*, 487, 10-15.

Gurr, M. I. and James, A. T. (1975). Lipid Biochemistry: An Introduction, 2nd ed. Chapman and Hall, London-John Wiley and Sons Inc., New York. 244 pp.

Hagerman, D., Wellington, F. D., and Villee, C. A. (1957). Estrogens in marine invertebrates. *Biol. Bull.*, 112, 180-183.

Harper, H. A. (1967). Review of Physiological Chemistry, 11th ed. Lange Medical Publications, Lost Altos, CA. 522 pp.

Helm, M. M., Holland, D. L., and Stephenson, R. R. (1973). The effect of supplementary algal feeding of a hatchery breeding stock of *Ostrea edulis* L. on larval vigour. *J. Mar. Biol. Ass. U.K.*, 53, 673-684.

Hemler, M. E. and Lands, W. E. M. (1977). Biosynthesis of prostaglandins. *Lipids*, 12, 591-595.

Hidu, H. and Tubiash, H. S. (1963). A bacterial basis for the growth of antibiotic - treated bivalve larvae. *Proc. Nat. Shellfish Assn.*, 54, 25-39.

Holland, D. L. and Gabbott, P. A. (1971). A microanalytical scheme for the determination of protein, carbohydrate, lipid and RNA levels in marine invertebrate larvae. *J. Mar. Biol. Ass. U.K.*, 51, 659-668.

Holland, D. L. and Spencer, B. E. (1973). Biochemical changes in fed and starved oysters, *Ostrea edulis* L. during larval development metamorphosis and early spat growth. *J. Mar. Biol. Assn. U.K.*, 53, 287-298.

Holland, D. L., Tantanasiriwong, R., and Hannant, P. J. (1975). Biochemical composition and energy reserves in larval and adults of the four British periwinkles *Littorina littorea*, *L. littoralis*, *L. saxatilis*, and *L. neritoides*. *Mar. Biol.* 33, 235-239.

Hopkins, G. J. and West, C. E. (1977). Diet induced changes in the fatty acid composition of mouse hepatocyte plasma membranes. *Lipids*, 12, 327-334.

Hoskin, G. P. and Hoskin, S. P. (1977). Partial characterization of the hemolymph lipids of *Mercenaria mercenaria* (Mollusca:Bivalvia) by thin-layer chromatography and analyses of serum fatty acids during starvation. *Biol. Bull.*, 152, 373-381.

Houtsmuller, U. M. T. (1972). Evaluation of modern foods as sources of lipids. *In:* "Lipids, Malnutrition and the Developing Brain". Ciba Foundation Symposium: 213-220. Assoc. Sci. Publishers, Amsterdam.

Hsu, C. M. L. and Kummerow, F. A. (1977). Influence of Elaidate and Erucate on heart mitochondria. *Lipids*, 12, 486-494.

Idler, D. R., Sangalang, G. B., and Kanazawa, A. (1969). Steroid desmolase in gonads of a marine invertebrate, *Placopecten magellanicus*. *Gen. Comp. Endocrinol.*, 12, 222-230.

Idler, D. R. and Wiseman, P. (1971). Sterols of Molluscs. *Int. J. of Biochemistry*, 2, 516-528.

Idler, D. R. and Wiseman, P. (1972). Molluscan sterols: a review. *J. Fish. Res. Bd. Can.*, 29, 385-398.

Imai, T. and Hatanaka, . (1949). On the artificial propagation of the Japanese common oyster, *Ostrea gigas* Thun., by non-colored naked flagellates. *Bull. Inst. Agric. Res.*, Tohoku Univ.,1, 33-46.

Isay, S. V., Makarchenko, M. A., and Vaskovsky, V. E. (1976). A study of glyceryl ethers - I. Content of glyceryl ethers in marine invertebrates from the sea of Japan and tropical regions of the Pacific ocean. *Comp. Biochem. Physiol.*, 55B, 301-305.

Kanazawa, A. and Teshima, S,-I. (1972). Sterols of the suspended matters in sea water. *J. Oceanogr. Soc. Jap.*, 27, 207-212.

Kanazawa, S., Teshima, S.-I., Ando, T., and Tomita, S. (1976). Sterols in coral reef animals. *Mar. Biol.*, 34, 532-557.

Karim, S. M. M. (1971a). Effects of oral administration of prostaglandin E_2 and $F_{2\alpha}$ on human uterus. *J. Obstet. Gynaec. Br. Commonw.*, 78, 289-293.

Karim, S. M. M. (1971b). Action of prostaglandin in the pregnant woman. *Ann. N. Y. Acad. Sci.*, 180, 483-498.

Karnaukhov, V. N., Milovidova, N. Y., and Kargopolova, I. N. (1977). On a role of carotenoids in tolerance of sea molluscs to environmental pollution. *Comp. Biochem. Physiol.*, 56A, 188-193.

Keenan, T. W. and Morre, D. J. (1970). Phospholipid class and fatty acid composition of Golgi apparatus isolated from rat liver and comparison with other cell fractions. *Biochemistry*, 9, 19–25.

Khalil, M. W. and Idler, D. R. (1976). Steroid biosynthesis in thw whelk *Buccinium undatum*. *Comp. Biochem. Physiol.*, 55B, 239–242.

Koganezawa, A. (1976). Mass rearing of offshore-type bivalve larval in Japan. *In:* "Proc. First Int. Conf. on Aquaculture Nutrition". pp. 162–171. (K. S. Price, Jr., W. N. Shaw, K. S. Danberg, eds.). Univ. of Delaware, Newark, Delaware.

Kok, L. T. and Norris, D. M. (1972). Comparative sterol compositions of adult female *Xyleborus ferrugineus* and its naturalistic fungal ectosymbionts. *Comp. Bio. Physio.*, 44, 499–505.

Kremer, B. P. and Schmitz, K. (1976). Aspects of $^{14}CO_2$-fixation by endosymbiotic rhodoplasts in the marine opisthobranchiate *Hermaea bifida*. *Mar. Biol.*, 34, 313–316.

Kuroda, M. and Endo, A. (1977). Inhibition of *in vitro* cholesterol synthesis by fatty acids. *Biochim. Biophys. Acta.*, 486, 70–81.

Lands, W. E. M., LeTellier, P. R., Rome, L., and Vanderhoek, J. Y. (1974). Regulation of prostaglandin synthesis. *In:* Prostaglandin Synthetase Inhibitors". pp. 1–7. (H. J. Robinson and J. R. Vane, eds.). Raven Press, New York, N.Y.

Larsson, Å. and Fänge, R. (1977). Cholesterol and free fatty acids (FFA) in the blood of marine fish. *Comp. Biochem. Physiol.*, 57B, 191–196.

Lawrence, J. M. (1965). "Lipid levels in body fluid, blood and tissues of some echinoderms and molluscs in relation to nutritional state". Ph.D. Thesis, Stanford University, Stanford, California.

Lawrence, J. M. (1976). Patterns of lipid storage in post-metamorphic marine invertebrates. *Amer. Zool.*, 16, 747–762.

Lee, F. O. and Cheng, T. C. (1971a). *Schistosoma mansoni* infection in *Biompalaria glabrata:* alterations in heart rate and thermal tolerance in the host. *J. Invert. Path.*, 18, 412–418.

Lee, F. O. and Cheng, T. C. (1971b). *Schistosoma mansoni:* respriometric and partial pressure studies in infected *Biomphalaria glabrata.* *Exp. Parasit.*, 30, 393–399.

Lee, F. O. and Cheng, T. C. (1972). Incorporation of ^{59}Fe in the snail *Biomphalaria glabrata* parasitized by *Schistosoma mansoni*. *J. Parasitol.*, 58, 481–488.

Lee, R. F., Nevenzel, J. C., and Paffenhofer, G.-A. (1971). Importance of wax esters and other lipids in the marine food chain: phytoplankton and copepods. *Marine Biol.*, 9, 99–108.

Lubet, P. E. and LeGall, P. (1974). Étude expérimentale de l'in-
    didence du cerveau sur le métabolisme des glucides et des
    lipides chey le mollusque mésogostropode. *(Crepidula fornicata*
    Phil.*)*. *Annales d'Endocrinologie* (Paris), 35, 383-386.
Lunetta, J. E. and Vernberg, W. B. (1971). Fatty acid composition
    of parasitized and non parasitized tissue of the mudflat
    snail, *Nassarius obsoletus* (Say). *Expl. Parasit.*, 30, 244-248.
Malek, E. A. and Cheng, T. C. (1974). Medical and Economic
    Malacology. Academic Press, New York. 398 pp.
Malins, D. C. and Wekell, J. C. (1969). The lipid biochemistry
    of marine organisms. *In:* "Progress in the chemistry of
    fats and other lipids." pp. 339-363. (R. T. Holman, ed.).
    Pergamon Press, Oxford.
Martoja, M. (1972). Endocrinology of mollusca. *Chem. Zool.*,
    7, 349-392.
Matsubosa, T. and Haugshi, A. (1973). Identification of molecu-
    lar species of ceramide aminoethylphosphonate from oyster
    adductor by gas-liquid chromatography-mass spectrometry.
    Biochim. Biophys. Acta., 296, 171-178.
McGandy, R. B. and Hegsted, D. M. (1975). Quantitative effects
    of dietary fat and cholesterol on serum cholesterol in man.
    In: "The role of fats in human nutrition". pp. 211-230.
    (A. J. Vergrosen, ed.). Academic Press, London.
McManus, D. P., Marshall, I., and James, B. L. (1975). Lipids
    in digestive gland of *Littorina saxatilis rudis* (Maton)
    and in daughter sporocysts of *Microphallus similis* (Jäg. 1900).
    Exptl. Parasit., 37, 157-163.
Millar, R. H. and Scott, J. M. (1967). The larvae of the oyster
    Ostrea edulis during starvation. *J. Mar. Biol. Ass. U.K.*,
    47, 475-484.
Muscatine, L. (1967). Glycerol excretion by symbiotic algae
    from corals and *Tridacna* and its control by the host. *Science*,
    156, 516-519.
Nestel, P. J., Havenstein, N., Homma, Y., Scott, T. W., and
    Cook, L. J. (1975). Increased sterol excretion with poly-
    unsaturated-fat high-cholesterol diets. *Metabolism*, 24, 189-
    198.
Odutuga, A. A. (1977). Recovery of brain from deficiency of
    essential fatty acids in rats. *Biochim. Biophys. Acta.*, 487,
    1-9.
Orten, J. M. and Newhaus, O. W. (1975). Human Biochemistry, 9th
    ed. C. V. Mosby Co., St. Louis, Missouri. 995 pp.
Oudejans, R. C. H. M. and van der Horst, D. J. (1974). Effect
    of excessive fatty acid ingestion upon composition of neutral
    lipids and phospholipids of snail *Helix pomatia* L. *Lipids*,
    9, 798-803.

Pardis, M. and Ackman, R. G. (1975). Occurrence and chemical structure of nonmethylene-interrupted dienoic fatty acids in American oyster *Crassostrea virginica*. *Lipids*, 10, 12-16.

Pardis, M. and Ackman, R. G. (1977). Potential for employing the distribution of anamolous nonmethylene-interrupted dienoic fatty acids in several marine invertebrates as part of food web studies. *Lipids*, 12, 170-176.

Patterson, G. W., Khalil, M. W., and Idler, D. R. (1975). Sterols of Scallop. I. Application of hydrophobic sephadex derivatives to the resolution of a complex mixture of marine sterols. *J. Chromat.*, 115, 153-159.

Piretti, M. V. and Viviani, R. (1976). Investigation of the constituent sterols of *Venus gallina*. *Comp. Biochem. Physiol.*, 55B, 229-234.

Porter, C. A. and Gamble, W. (1973). Observations on the effect of the rediae of *Nanophyetus salmincola* on the fatty acid content of the hepatopancreas of *Oxytrema silicula* (Gould). *Comp. Biochem. Physiol.*, 45B, 905-909.

Rahm, J. J. and Holman, R. T. (1964). Studies of the metabolism of polyunsaturated acids by short-term experiments. *J. Nutrition*, 84, 149-154.

Rapport, M. M. (1961). The α,β unsaturated ether (plasmologen) content of the tissues of several molluscs. *Biol. Bull.*, 121, 376-377.

Rapport, M. M. and Alonzo, N. F. (1960). The structure of plasmologens. V. Lipids of marine invertebrates. *J. Biol. Chem.*, 235, 1953-1956.

Rossiter, R. J. and Strickland, . (1960). The metabolism and function of phosphatides. *In:* "Lipide Metabolism". pp. 69-127. (K. Bloch, ed.). John Wiley & Sons, Inc., New York.

Rathbone, L. (1965). The effect of diet on the fatty acid compositions of serum, brain, brain mitochondria and myelin in the rat. *Biochem. J.*, 97, 620-628.

Runham, N. W. (1975). Alimentary Canal. *In:* "*Pulmonates. I:* Functional Anatomy and Physiology". pp. 53-104. (V. Fretter and J. Peake, eds.). Academic Press, New York.

Salaque, A., Barbier, M., and Lederer, E. (1966). Sur la biosynthése des sterols de l'huitre *Ostrea gryphea* et de l'oursin *Paracentrotus lividus*. *Comp. Biochem. Physiol.*, 19, 45-51.

Saliot, A. and Barbier, M. (1971). Sur l'isolement de la progésterone et de quelques cétostéroides de las partie femelle des gonades de la Coquille Saint-Jacques *Pecten maximus*. *Biochimie*, 53, 265-266.

Sherman, H. C. (1941). "Chemistry of Food and Nutrition", 6th ed. The MacMillan Co., New York. 611 pp.

Slowey, J. F., Jeffrey, L. M., and Hood, D. W. (1962). The fatty acid content of ocean water. *Geochim. et Cosmochim. Acta.*, 26, 607-616.

Smith, W. L. and Chanley, M. H. (1975). Culture of Marine Invertebrates. Plenum Press, New York. 338 pp.

Snyder, R. (1969a). The biochemistry of lipids containing ether bonds. *In:* "Progress in the chemistry of fats and other lipids". pp. 287-335. 10 (R. T. Holman, ed.). Pergamon Press, Oxford.

Snyder, F. (1969b). Ethers-linked lipids in neoplasms of man and animals: methods of measurement and the occurrence and nature of the alkyl and alk-1-enyl moieties. *In:* "Advances in experimental Medicine and Biology". pp. 609-621. 4, Plenum Press, New York.

Snyder, F. (1972). Enzyme systems that synthesize and degrade glycerolipids possessing ether bonds. *Adv. in Lipid Res.,* 10, 233-259.

Standen, O. D. (1951). Some observations upon the maintenance of *Australorbis glabratus* in the laboratory. *Ann. Trop. Med. Parasitol.,* 45, 80-83.

Stern, G. (1970). Production et bilan energetique chez la limace rouge. *Terre Vie,* 24, 403-424.

Steudler, P. A., Schmitz, F. J., and Ciereszko, L. S. (1977). Chemistry of coelenterates. Sterol composition of some predator-prey pairs on coral reefs. *Comp. Biochem. Physiol.,* 56B, 385-392.

Sugita, M., Itasaka, O., and Hori, T. (1976). Branched long-chain bases from the bivalve *Corbicula sandai*. *Chem. Physic. Lipids,* 16, 1-8.

Sumner, A. T. (1965). Experiments on phagocytosis and lipid absorption in the alimentary system of *Helix*. *J. Roy. Mic. Soc.,* 84, 415-421.

Suryanarayanan, H. and Alexander, K. M. (1971). Fuel reserves of molluscan muscle. *Comp. Biochem. Physiol.,* 40A, 55-60.

Suryanarayanan, H. and Alexander, K. M. (1973). Biochemical studies on red muscles of the gastropod *Pila virens,* with a note on its histochemistry. *Comp. Biochem. Physiol.,* 44A, 1157-1162.

Tamura, T., Truscott, B., and Idler, D. R. (1964). Sterol metabolism in the oyster. *J. Fish Res. Bd. Canada,* 21, 1519-1522.

Taylor, D. L. (1971). Photosynthesis of symbiotic chloroplasts in *Tridachia crispata* (Bergh). *Comp. Biochem. Physiol.,* 38A, 233-236.

Taylor, D. L. (1973a). The cellular interactions of algal-invertebrate symbiosis. *Adv. Mar. Biol.,* 2, 1-56.

Taylor, D. L. (1973b). Algal symbionts of invertebrates. *Ann. Rev. Microbiol.,* 27, 171-187.

Teshima, S.-I. and Kanazawa, A. (1973). Biosynthesis of 7-cholestenol in the chiton, *Liolophura japonica*. *Comp. Biochem. Physiol.,* 44B, 881-887.

Teshima, S.-I. and Kanazawa, A. (1974). Biosynthesis of sterols in abalone, *Haliotis-gurneri* and mussel, *Mytilus edulis*. *Comp. Biochem. Physiol.*, 47B, 555-561.

Thiele, O. W. (1959). Der lipide de Weinbergschnecke (*Helix pomatia* L.) I. Mitteilung: jahreszeitliche veranderungen in der zussammensetzung der lipide. *Z. Vergl. Physiol.*, 42, 484-491.

Thompson, G. A., Jr. (1966). The biosynthesis of ether-containing phospholipids in the slug, *Arion ater*. II. The role of glycerol ether lipids as plasmologen precursors. *Biochemistry*, 5, 1290-1296.

Thompson, G. A., Jr. (1968). The biosynthesis of ether-containing phospholipids in the slug, *Arion ater*. III. Origin of the vinylic ether bond of plasmologens. *Biochem. Biophysics. Acta.*, 152, 409-411.

Thompson, G. A., Jr. (1972a). Ether-linked lipids in molluscs. *In:* "Ether lipids: Chemistry and Biology". pp. 313-320. (F. Snyder, ed.). Academic Press, New York.

Thompson, G. A., Jr. (1972b). Ether-linked lipids in protozoa. *In:* "Ether lipids: Chemistry and Biology". pp. 321-327. (F. Snyder, ed.). Academic Press, New York.

Thompson, R. J., Ratcliffe, N. A., and Bayne, B. L. (1974). Effects of starvation on structure and function in the digestive gland of the mussel (*Mytilus edulis* L.). *J. Mar. Biol. Ass. U.K.*, 54, 699-712.

Thompson, S. N. (1973). A review and comparative characterization of the fatty acid compositions of seven insect orders. *Comp. Biochem. Physiol.*, 45B, 467-482.

Tinoco, J., Miljanich, P., and Medwadowski, B. (1977). Depletion of docosahexaenoic acid in retinal lipids of rats fed a linolenic acid-deficient, linoleic acid-containing diet. *Biochim. Biophys. Acta.*, 486, 575-578.

Tornabene, T. G., Kates, M., and Volcani, B. E. (1974). Sterols alaphatic hydrocarbons, and fatty acids of a nonphotosynthetic diatom, *Nitschia alba*. *Lipids*, 9, 279-284.

Treguer, P., LeCorre, P., and Courtot, P. (1972). A method for determination of the total dissolved free fatty-acid content of sea water. *J. Mar. Biol. Ass. U.K.*, 52, 1045-1055.

Ukeles, R. (1976). Views on bivalve larvae nutrition. *In:* "Proc. First Int. Conf. on Aquaculture Nutrition". pp. 127-162. (K. S. Price, Jr., W. N. Shaw, D. S. Danberg, eds.). Univ. of Delaware, Newark, Delaware.

van der Horst, D. J. (1973). Biosynthesis of saturated and unsaturated fatty acids in the pulmonate land snail *Cepaea nemoralis* (L.). *Comp. Biochem. Physiol.*, 46B, 551-560.

van der Horst, D. J. (1974). *In vivo* biosynthesis of fatty acids in the pulmonate land snail *Cepaea nemoralis* (L.) under anoxic conditions. *Comp. Biochem. Physiol.*, 47B, 181-187.

van der Horst, D. J., Kingma, F. J., and Oudejans, R. C. H. M. (1973). Phospholipids of the pulmonate land snail *Cepaea nemoralis* (L.). *Lipids*, 8, 759-765.

van der Horst, D. J., Oudejans, R. C. H. M., Meijers, J. A., and Testerink, G. J. (1974). Fatty acid metabolism in hibernating *Cepaea nemoralis* (Mollusca:Pulmonata). *J. Comp. Physiol.*, 91, 247-256.

van der Horst, D. J. and Oudejans, R. C. H. M. (1976). Fate of dietary linoleic and linolenic acids in the land snail *Cepaea nemoralis* (L.). *Comp. Biochem. Physiol.*, 55B, 167-170.

van der Horst, D. J. and Zandee, D. I. (1973). Invariability of the composition of fatty acids and other lipids in the pulmonate land snail *Cepaea nemoralis* (L.) during an annual cycle. *J. Comp. Physiol.*, 85, 317-326.

von Brand, T., McMahon, P., and Noland, M. (1957). Physiological observations on starvation and desiccation of the snail. *Australorbis glabratus*. *Bio. Bull.*, 113, 89-102.

Voogt, P. A. (1967a). Biosynthesis of 3β-sterols in a snail, *Arion rufus* L., from 1-^{14}C-acetate. *Archs. Int. Physiol. Biochem.*, 75, 492-500.

Voogt, P. A. (1967b). Investigations on the capacity of synthesizing 3β-sterols in Mollusca. I. Absence of 3β-sterol synthesis in a whelk, *Buccinium undatum* L. *Arch. Int. Physiol. Biochem.*, 75, 809-815.

Voogt, P. A. (1968a). Investigations of the capacity of synthesizing 3-β-sterols in Mollusca. II. Study on the biosynthesis of 3β-sterols in some representatives of the order Basommatomorpha. *Comp. Biochem. Physiol.*, 25, 943-948.

Voogt, P. A. (1968b). Investigations of the capacity of synthesizing 3β-sterols in Mollusca. III. The biosynthesis of 3β-sterols in some archeogastropods. *Archs. Int. Physiol. Biochem.*, 76, 721-730.

Voogt, P. A. (1969). Investigations of the capacity of synthesizing 3β-sterols in Mollusca. IV. The biosynthesis of 3β-sterols in some mesogastropods. *Comp. Biochem. Physiol.*, 31, 37-46.

Voogt, P. A. (1971). Investigations of the capacity of synthesizing 3β-sterols in mollusca. V. The biosynthesis and composition of 3β-sterols in the mesogastropods *Crepidula fornacata* and *Natica cataena*. *Comp. Biochem. Physiol.*, 39B, 139-149.

Voogt, P. A. (1972). Lipid and sterol components and metabolism in mollusca. *Chem. Zool.*, 7, 245-300.

Voogt, P. A. (1975a). Investigations of the capacity of synthesizing 3β-sterols in mollusca. XIII. Biosynthesis and composition of sterols of some bivalves (Anisomyaria). *Comp. Biochem. Physiol.*, 50B, 499-504.

Voogt, P. A. (1975b). Investigations of the capacity of synthesizing 3β-sterols in mollusca. XIV. Biosynthesis and composition of sterols of some bivalves (Eulamellibranchia). *Comp. Biochem. Physiol.*, 50B, 506–510.

Walton, M. J. and Pennock, J. F. (1972). Some studies on the biosynthesis of ubiquinone, isoprenoid alcohols, squalene and sterols by marine invertebrates. *Biochem. J.*, 127, 471–479.

Watanabe, T. and Ackman, R. G. (1972). Effect of unicellular algal lipids on oyster fatty acid composition. *Fish. Res. Bd. Can. Tech. Rept.* No. 334.

Watanabe, T. and Ackman, R. G. (1974). Lipids and fatty acids of the American *(Crassostrea virginica)* and European flat *(Ostrea edulis)* oysters from a common habitat, and after one feeding with *Dicrateria inornata* or *Isochrysis galbana*. *J. Fish. Res. Bd. Can.*, 31, 403–409.

Watanabe, T. and Ackman, R. G. (1977). Effect of storage on lipids and fatty acids of oysters. *J. Inst. Can. Sci. Technol. Aliment.*, 10, 40–42.

Williams, C. S. (1969). The effect of *Mytilicola intestinalis* on the biochemical composition of mussels. *J. Mar. Biol. Ass. U.K.*, 49, 161–173.

Williams, E. E. (1970). Seasonal variations in the biochemical composition of the edible winkle *Littorina littorea* (L.). *Comp. Biochem. Physiol.*, 33, 655–661.

Williams, P. M. (1965). Fatty acids derived from lipids of marine origin. *J. Fish. Res. Bd. Can.*, 22, 1107–1122.

Wolfe, D. A., Rao., P. V., and Cornwell, D. G. (1965). The fatty acid composition of crayfish lipids. *J. Am. Oil Chem. Soc.*, 42, 633–637.

Zandee, D. I. (1967). Absence of cholesterol synthesis in *Sepia officinalis*. *Arch. Int. Physiol. Biochem.*, 75, 487–491.

Zs-Nagy, I. (1971). The lipochrome pigment of molluscan neurons as a specific electron acceptor. *Comp. Biochem. Physiol.*, 40A, 595–602.

Zs-Nagy, I. and Ermini, M. (1972). Oxidation of $NADH_2$ by the lipochrome pigment of the tissues of the bivalve *Mytilus galloporvincialis* (Mollusca, Pelecypoda). *Comp. Biochem. Physiol.*, 43, 39–46.

The Role of Lysosomal Hydrolases in Molluscan Cellular Response to Immunologic Challenge[1]

THOMAS C. CHENG

Institute for Pathobiology
Center for Health Sciences
Lehigh University
Bethlehem, Pennsylvania

[1]This research was supported by a grant (AI 12355-03) from the National Institute of Allergy and Infectious Diseases, and a grant (FD 00416-05) from the Food & Drug Administration, U.S. Public Health Service.

I. INTRODUCTION

Although considerable new information has been contributed in
recent years to our knowledge of the biochemistry and physiology
of phagocytosis by invertebrate hemocytes, especially those of
insects and molluscs (see Cheng, 1975, and Anderson, 1977, for
reviews), little new information has been contributed to our
understanding of what is generally considered a second type of
cellular reaction in molluscs to invading nonself materials,
i.e., encapsulation or granuloma formation. Since the compre-
hensive review of what is known about this process and related
phenomena in molluscs by Cheng and Rifkin (1970), new information
has been contributed only by Harris and Cheng (1975a,b), Harris
(1975), Rachford (1976), Krupa *et al.* (1977), Cheng and Garrabrant (1977), and Cheng (1978).

By employing the nematode *Angiostrongylus cantonensis* and the
pulmonate gastropod *Biomphalaria glabrata* as the experimental
model, Harris and Cheng (1975a) have reported that the encapsulation of the helminth parasite in the mollusc commences at 24-48 hr
post-infection. Furthermore, they found that granuloma formation
occurs in two phases: (1) initial infiltration and aggregation
of hemocytes around the parasite, and (2) subsequent transformation
into a more fibrous-appearing granuloma. Subsequent study of the
composition of such granulomas by transmission electron microscopy
by Harris (1975) has revealed that the "fibrous-appearing" granuloma was in fact comprised of long and thin cytoplasmic extension
of granulocytes. I concur with the opinion of Harris (1975) that
what have been referred to as "fibroblasts" by earlier investigators (Newton, 1952, 1954; Barbosa and Barreto, 1960; Tripp, 1961;
Pan, 1963, 1965), who have studied similar granulomas with the
light microscope, are actually granulocytes at various stages
of extending pseudopodia.

Rachford (1976), who studied the cellular reactions in *Lymnaea
palustris* to *A. cantonensis,* reported that the granulomas were
comprised of amoebocytes, fibroblasts, and pigment cells. Krupa
et al. (1977) studied the cellular reaction of *Bulinus guernei* to
sporocysts of the trematode *Schistosoma haematobium* at the
electron microscope level. Since the strain of *B. guernei* which
they employed originated from Gambia and was not totally compatible with the Egyptian strain of *S. haematobium* used, there was
some cellular reaction in the molluscan host. The resulting
granulomas were comprised of both granulocytes and hyalinocytes,
but only the granulocytes came in contact with the microvilli on
the sporocyst surfaces. Furthermore, they confirmed Harris's
(1975) finding that these granulocytes produced fine filopods
which formed a multilamellated encapsulation complex around the
parasites.

The only studies to date which have given some clue to the chemical nature of molluscan granulomas are those of Harris and Cheng (1975a), Cheng and Garrabrant (1977), and Cheng (1978). Harris and Cheng, by employing enzyme histochemistry, have reported that at 4 weeks after penetration of *B. glabrata* by *A. cantonensis* larvae, there were highly localized acid phosphatase and nonspecific esterase activities associated with each granuloma, and critical examination by me of Harris's slides has revealed that although these hydrolases were primarily limited to within granulocytes, some had diffused to the intercellular spaces. In addition to these enzymes, lesser amounts of alkaline phosphatase and β-glucuronidase activities also occurred within each granuloma, but aminopeptidase activity could not be demonstrated. In addition, it was ascertained that the acid phosphatase activity gradually increased as the encapsulation reaction progressed from the first to the fourth week post-penetration by *A. cantonensis*.

Cheng and Garrabrant (1977) demonstrated that the granulomas formed in totally and partially incompatible strains of *B. glabrata* in response to *S. mansoni* larvae were comprised totally of granulocytes and these cells, as Harris (1975) had shown, were rich in acid phosphatase. In fact, since it had been demonstrated by studying isolated cells that this enzyme only occurred in granulocytes and not in hyalinocytes, Cheng and Garrabrant concluded that acid phosphatase was a useful marker to trace the role of granulocytes in granuloma formation. It is also important to note that Cheng and Garrabrant (1977) and Cheng (1978) have noted that there was an increase in intracellular acid phosphatase activity in older granulomas. In brief, it now appears that granulomas formed in molluscs in response to helminth parasitism are primarily, if not exclusively, comprised of granulocytes. Furthermore, these granulocytes include acid phosphatase and other lysosomal hydrolases, e.g., esterase, alkaline phosphatase, and β-glucuronidase. The question that now must be asked is: Do these hydrolases always contribute to the destruction of the encapsulated parasite? Prior to presenting some experimental evidence and speculations, another important point should be remembered. This is considered below.

Cheng (1967) and Harris and Cheng (1975b) pointed out that granulomas formed in response to helminth parasites in molluscs need not always lead to their destruction. For example, the encapsulation of the nematode *Angiostrongylus cantonensis* in *Achatina fulica, Biomphalaria glabrata, Lymnaea palustris,* and others does not lead to the death of the parasite (Cheng and Rifkin, 1970; Harris and Cheng, 1975a; Rachford, 1976). On the other hand, although it still remains uncertain what are the exact causes of the death and subsequent destruction of certain encapsulated helminths, such as *S. mansoni* mother sporocysts in

incompatible strains of *Biomphalaria glabrata* (see Cheng and
Garrabrant, 1977), it is noted that Sminia (1972) reported the
hypersynthesis of acid phosphatase in hemocytes of *Lymnaea
stagnalis* after *in vitro* challenge with bacteria. Also, Cheng
(1976) presented a summary of earlier studies which indicated
the hypersynthesis of several other lysosomal enzymes in molluscan hemocytes and their subsequent release into serum as a
result of challenge by nonself substances. He also pointed
out that the released enzymes were capable of lysing several
species of bacteria. The questions that have to be asked are:
Do the elevated levels of lysosomal hydrolases in the serum of
challenged molluscs contribute to the destruction of at least
certain metazoan parasites? and why are certain helminths, such
as *Angiostrongylus cantonensis*, not destroyed within granulomas,
while others, such as *Schistosoma mansoni* sporocysts, are killed
and degraded? Presented below are some experimental evidences
for the action of elevated serum hydrolases on bacteria as
determined at the electron microscope level and some thoughts
as to why certain helminths are not destroyed.

II. MATERIALS AND METHODS

A. *BACTERIAL INJECTIONS AND CONTROLS*

The specimens of *Biomphalaria glabrata* employed were of the
so-called NIH albino strain (Newton, 1955). Each measured 6-8
mm in shell diameter. All of the snails were maintained as
previously described (Harris and Cheng, 1975b).

Snails of the experimental group, consisting of 40 specimens,
were divided into two equal groups. Each snail of one group was
injected through the shell (into the visceral region) with 5 µl
of aqueous suspension of heat-killed *Bacillus megaterium* at a
concentration of 80×10^6 bacteria/ml. The members of the second
group were injected with a same amount of bacterial suspension
into the sole of the foot.

One set of controls consisting of 40 snails were sham-injected,
with 20 punctured in the visceral mass and the remaining 20 in
the foot. A second set of controls consisting of 20 snails was
not manipulated.

B. *PREPARATION OF SERUM*

Two hours after the snails of the experimental group were
challenged, hemolymph samples were collected intracardially
from each by the method of Cheng and Yoshino (1976). Similarly,
hemolymph samples were collected from the two groups of control
snails. The samples from each category of snails were pooled
and subjected to centrifugation at 400 g for 10 min at -5° C in

a Sorvall RC2-B refrigerated centrifuge to separate the cells from serum.

After the serum had been decanted from the cells, 20×10^6 live vegetative cells of B. *megaterium* that had been thoroughly washed were placed in each category of serum. In addition, an identical concentration of fresh B. *megaterium* was placed in a 0.023 g/liter solution of pure lysozyme (Muramidase, Sigma, St. Louis, Missouri). Also, a same concentration of live B. *megaterium* was placed in a 0.07 M NaCl solution.

All of the bacterial suspensions were exposed to the respective sera, enzyme, or salt solutions for 8 hr with intermittent agitation. After this exposure period, each preparation was centrifuged at 400 g for 10 min at -5° C to pellet the bacterial cells.

C. *ELECTRON MICROSCOPY*

The bacteria were then fixed with 0.075 M cacodylate-buffered 6% glutaraldehyde, pH 7.45, for 1 hr, followed by several washings in cold buffer for a total of 2 hr. The bacteria were subsequently fixed in 1% osmium tetroxide in cacodylate buffer for 3 hr, starting at 5° C but permitting it to warm up to room temperature during the fixation process. This was followed by washing in cold buffer for four additional times, 5 min each, followed by dehydration in an ascending ethanol series. Propylene oxide was employed as the final dehydrant at room temperature for two 20-min washings.

The cells were subsequently embedded in a 1:1 mixture of Spur's Epon and propylene oxide, and eventually transferred to pure Epon. Subsequently, the preparation was centrifuged at 800 g for 10 min to form a cell pellet, which was permitted to polymerize at 70° C for 10 hr. The sections were cut on a Porter-Blum MT-2 ultramicrotome, stained for 15 min with uranyl acetate, followed by 1 min with lead citrate. They were examined in a Hitachi HS-8 electron microscope.

III. RESULTS

A. *UNCHALLENGED CONTROLS*

Vegetative cells of B. *megaterium* that had been exposed to 0.07 M NaCl solution showed no abnormalities (Fig. 1). Their walls are intact as is the enclosed protoplasm. Certain cells portray more or less completely developed centripetally oriented transverse cell walls (Fig. 1) as well as low-density fibrous component of the nuclear apparatus surrounding dense nuclear material (Fig. 1).

Fig. 1. Electron micrograph of *Bacillus megaterium* that had
         been exposed for 8 hr to 0.07 M NaCl. Note unaltered
         cell wall (w), centripetal thickened cell wall (cw) as
         an indication of division, dense nuclear apparatus (DN),
         and normal appearing cytoplasm. x 33,000.

Fig. 2. Electron micrograph of *B. megaterium* that had been
         exposed for 8 hr to serum of sham-injected *Biomphalaria
         glabrata*. Note unaltered cell wall and essentially
         normal appearance of cell. x 30,000.

Fig. 3. Electron micrograph of *B. megaterium* that had been
         exposed for 8 hr to serum of untampered *B. glabrata*.
         Note essentially normal appearance of cell. x 30,000.

Fig. 4. Electron micrograph of *B. megaterium* that had been
         exposed for 8 hr to serum of *B. glabrata* which had
         been previously challenged with *B. megaterium*. Note
         eroded wall (w) and totally disrupted inclusions.
         x 26,000.

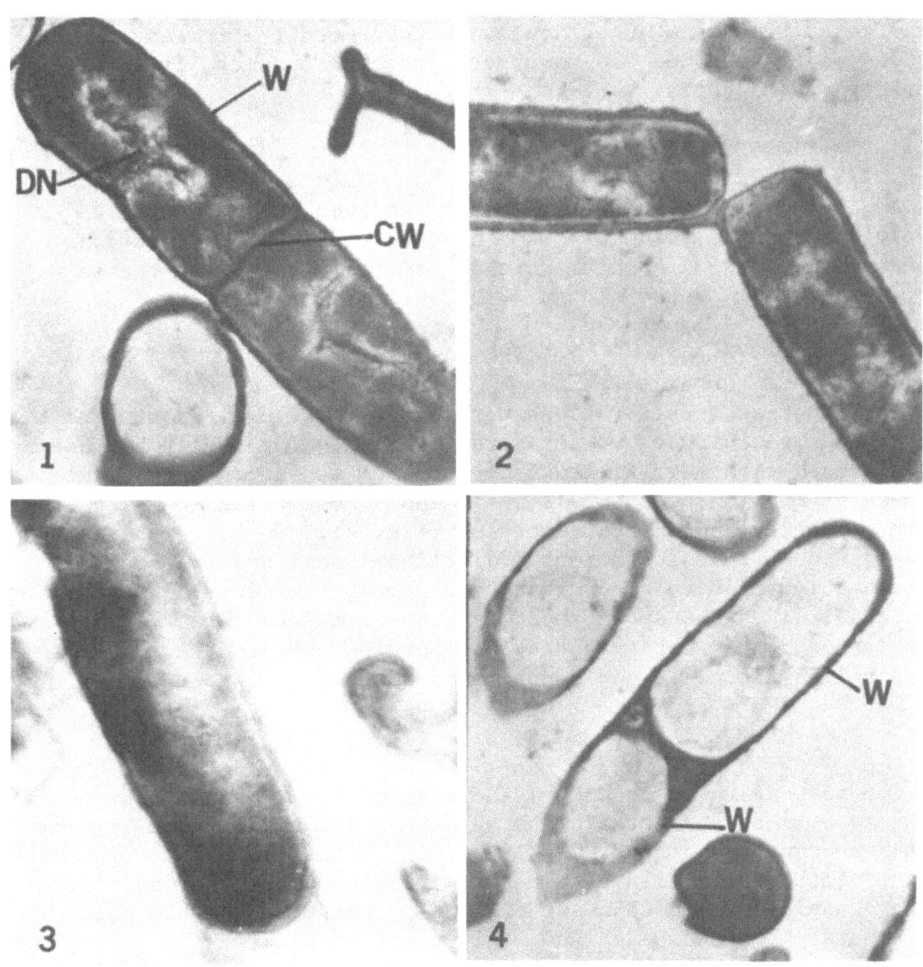

B. *SHAM-INJECTED CONTROLS*

B. *megaterium* that had been exposed to the sera of sham-injected
snails are essentially normal in appearance at the EM level
(Fig. 2). The cell wall is intact as is the cytoplasm. This
indicates that sham-injection had not induced the release of
sufficient lysozyme, and/or other lysosomal enzymes, if any, to
affect the bacteria.

C. *UNTAMPERED CONTROLS*

As is the case with bacteria that had been exposed to salt
solution and the serum of sham-injected snails, no alterations
were noticed in B. *megaterium* that had been exposed to the serum
of untampered B. *glabrata* (Fig. 3).

D. *EXPERIMENTALS*

Rather dramatic alterations were observed with B. *megaterium*
that had been exposed to the serum of B. *glabrata* which had been
challenged with heat-killed B. *megaterium in vivo*. Specifically,
there is erosion of the cell wall, and almost total leaching of
the cytoplasm and nuclear material (Fig. 4). In some instances
where a centripetally oriented divisional cell wall had formed,
presumably prior to exposure to the serum, the constituent
material gives the appearance of having remained intact (Fig. 4).
It is noted that bacteria that had been exposed to the serum of
prechallenged snails could not be recultured on nutrient agar.

E. *LYSOZYME-TREATED CONTROLS*

The most dramatic changes associated with B *megaterium* were
observed in those bacteria that had been exposed to pure lysozyme
at a concentration of 0.023 g/liter for 8 hr. In these, although
much of the cell walls retained their shape, enzyme-eroded areas
are conspicious and abundant (Fig. 5). Furthermore, the cell
contents, both cytoplasm and nuclear material, are no longer
apparent. In their place, the intracellular spaces are lucid
and empty (Fig. 5). Bacteria that had been exposed to lysozyme
could not be recultured on nutrient agar.

IV. DISCUSSION AND CONCLUSIONS

Earlier studies on B. *glabrata* (Cheng and Yoshino, 1976, Cheng
et al., 1977, 1978) have indicated that actively phagocytosing
hemocytes undergo hypersynthesis of at least certain lysosomal
enzymes, and some of these enzymes are released into serum.
Specifically, it has been demonstrated that when challenged with
live or heat-killed *Bacillus megaterium*, there are elevations in

both intracellular lipase and its subsequent release (Cheng and Yoshino, 1976), the hypersynthesis and release of lysozyme (Cheng *et al.*, 1977), as well as the hypersynthesis and release of aminopeptidase (Cheng *et al.*, 1978).

The question being asked is: Do the elevated levels of lysosomal enzymes have any protective effect against invading organisms? It is noted that McDade and Tripp (1967) demonstrated that lysozyme in the hemolymph of the oyster, *Crassostrea virginica*, has an antimicrobial effect, causing the lysis of certain Gram-positive bacteria such as *Micrococcus lysodeikticus*, *Bacillus megaterium*, and *Bacillus subtilis*. In addition, Rodrick and Cheng (1974) demonstrated that the lysozyme from the same mollusc will lyse *Escherichia coli*, *Gaffkya (=Aerococcus) tetragena*, *Salmonella pullorum*, and *Shigella sonnei* in addition to *B. subtilis* and *B. megaterium*. These studies were all carried out by the conventional sensidisc method or by biochemically assaying for breakdown products.

From the data presented, it is apparent that the elevated levels of certain lysosomal enzymes in molluscan serum resulting from intracellular hypersynthesis due to exogenous challenge are sufficient to affect bacteria, causing their death and degradation. This is certainly the case with *B. megaterium*. The specificity of such enzymatic degradation has not been ascertained, although there is little doubt that it is not as specific as the immunogen-immunoglobulin complex of vertebrates (Cheng *et al.*, 1978). Nevertheless, there can be no doubt that microorganisms that are susceptible to lysosomal enzymes, as indicated by the *in vitro* studies reported, are degraded by the elevated levels of lysosomal enzymes resulting from foreign material challenge.

V. SOME THEORETICAL CONSIDERATIONS

Because of the nature of this symposium, I am taking the liberty of sharing some hypotheses with you. Specifically, I wish to direct my remarks at the possible roles of lysosomal enzymes that occur in molluscan hemocytes relative to metazoan parasites.

As a result of the studies by McDade and Tripp (1967), Rodrick and Cheng (1974), Cheng and Rodrick (1974), and the data presented herein, a pattern is evolving, i.e., elevated levels of lysosomal enzymes in molluscan serum can serve as defense molecules against susceptible microorganisms. Since it has been suggested (Cheng and Rifkin, 1970) that encapsulation of helminth parasites, or granuloma formation, actually represents aborted attempts at phagocytosis of foreign particles too large to be engulfed, the question should be raised as to what mechanisms underlie the fact that certain encapsulated metazoan parasites, such as incompatible schistosome larvae, are destroyed, while others, such as *Angiostrongylus cantonensis*, are not? The hypothetical model, illustrated in Figure 6, illustrates my idea.

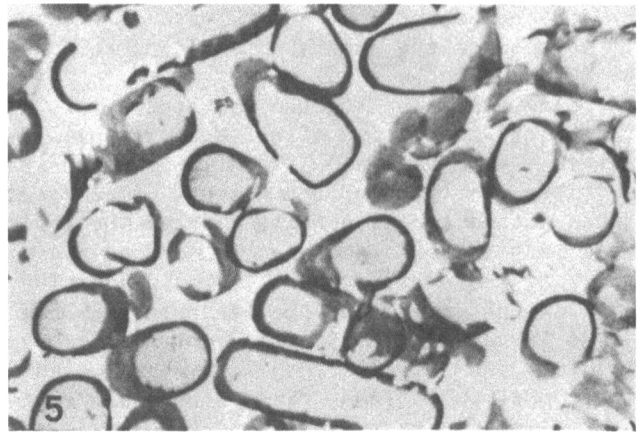

Fig. 5.   Electron micrograph of *Bacillus megaterium* that had been
          exposed to 0.23 g/liter of pure lysozyme. Note dis-
          ruption of cell walls and empty inclusions. x 15,000.

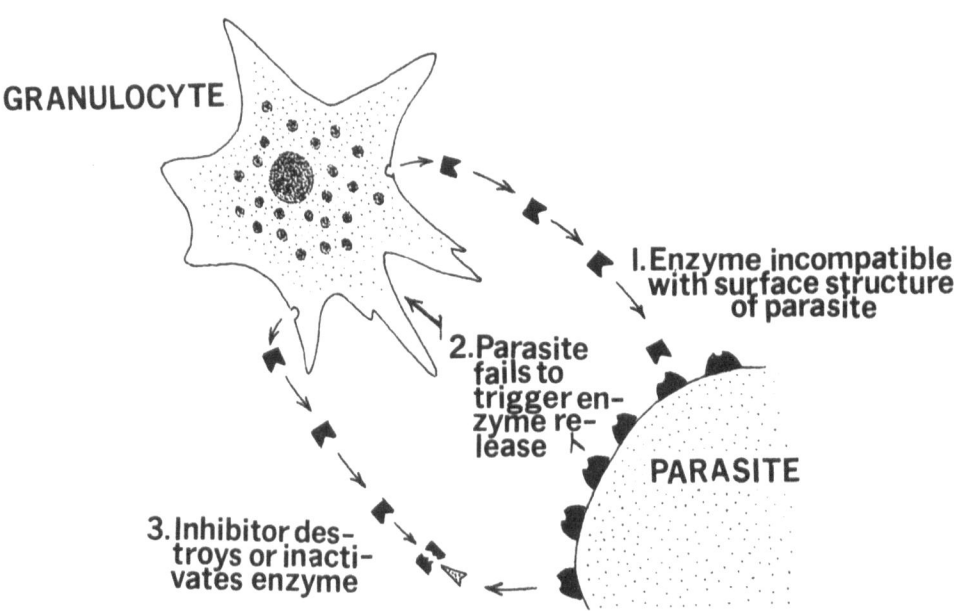

Fig. 6.   Schematic diagram illustrating possible mechanisms which
          may be responsible for the nondestruction of metazoan
          parasites by lysosomal enzymes from molluscan granulo-
          cytes.

It is known as a result of the work of Harris and Cheng (1975b), Cheng and Garrabrant (1977), and Cheng (1978) that there is an accumulation of acid phosphatase in the granulocytes comprising the parasite-encapsulating granuloma. Phosphatases are hydrolytic enzymes which catalyze the hydrolysis of phosphoric esters:

$$\text{Glycerol 1 - phosphate} + H_2O \rightarrow \text{glycerol} + P_i$$

Acid phosphatase, a widely recognized lysosomal marker enzyme, hydrolyzes a variety of phosphoric esters at an acid pH. As state earlier, there is diffusion of this enzyme from granuloma-comprising capsules surrounding *A. cantonensis* larvae, yet the parasite is unaffected. This suggests that the susceptible phosphoric esters, at least in the exposed configuration, are not available and hence the hydrolase has no deleterious effect. Relative to this thesis, it should be recalled that Ubelaker *et al.* (1970) have found that when hemocytes (granulocytes) of the beetle *Tribolium confusum* make contact with the cysticercoids of the cestode *Hymenolepis diminuta,* some of the host cells are ruptured, undoubtedly releasing enzymes; yet the helminths are not damaged. This, again, may be an example of the absence of a suitable substrate.

Thus one postulation as to why elevated serum hydrolases do not affect certain nonself substances is the absence of vulnerable substrates on their surfaces.

The second possibility is that the release of lysosomal enzymes from their sites of synthesis, i.e., granulocytes and other sources (Yoshino and Cheng, 1977), must be triggered by some component of the parasite's somatic antigen and/or some secretion. If such a triggering molecule(s) is absent, then one would not expect the release of antagonistic enzymes, at least to deleterious levels, and the encapsulated parasite is not affected.

A third possibility exists. It is possible that antienzymes are elaborated by the encapsulated parasite which inactivates the lytic enzyme. This remains to be investigated.

In summary, there is now a body of evidence that there are "humoral" protective molecules in molluscs which may have direct protective functions, but these are not chemically immunoglobulins or opsonins. These are lysosomal enzymes which are limited in their specificity. Of course, there are other soluble molecules, such as agglutinins, which are known to play a role in enhancing phagocytosis.

ACKNOWLEDGMENTS

I wish to acknowledge the technical assistance of Douglas R. Keene during the preliminary stages of this study, and to my son, J. Bradford Cheng, for his expert assistance with photography.

REFERENCES

Anderson, R. S. (1977). Immune mechanisms in invertebrates.
    Comp. Pathobiol., 3, 1-20.
Barbosa, F. S. and Barreto, A. C. (1960). Differences in sus-
    ceptibility of Brazilian strains of *Australorbis glabratus*
    to *Schistosoma mansoni*. *Exptl. Parasitol.*, 9, 137-140.
Cheng, T. C. (1967). Marine molluscs as hosts for symbioses:
    with a review of known parasites of commercially important
    species. *Adv. Mar. Biol.*, 5, 1-424.
Cheng, T. C. (1975). Functional morphology and biochemistry of
    molluscan phagocytes. *Ann. N. Y. Acad. Sci.*, 266, 343-379.
Cheng, T. C. (1976). Humoral immunity in molluscs. Proc. 1st
    Internat. Colloq. Invert. Pathol., pp. 190-194. Queen's
    Univ. Press, Kingston, Canada.
Cheng, T. C. (1978). A study of granuloma formation by molluscan
    cells. *Comp. Pathobiol., This volume.*
Cheng, T. C. and Garrabrant, T. A. (1977). Acid phosphatase in
    granulocytic capsules formed in strains of *Biomphalaria
    glabrata* totally and partially resistant to *Schistosoma
    mansoni*. *Intl. J. Parasitol.*, 7, 467-472.
Cheng, T. C. and Rifkin, E. (1970). Cellular reactions in marine
    molluscs in response to helminth parasitism. *In:* "A
    Symposium on Diseases of Fishes and Shellfishes." (S. F.
    Sniesko, ed.). pp. 443-496. *Am. Fisher. Soc.*, Spec. Publ.
    No. 5, Washington, D.C.
Cheng, T. C. and Rodrick, G. E. (1974). Identification and
    characterization of lysozyme from the hemolymph of the soft-
    shelled clam *Mya arenaria*. *Biol. Bull.*, 147, 311-320.
Cheng, T. C. and Yoshino, T. P. (1976). Lipase activity in the
    hemolymph of *Biomphalaria glabrata* (Mollusca) challenged with
    bacterial lipids. *J. Invert. Pathol.*, 28, 143-146.
Cheng, T. C., Chorney, M. J., and Yoshino, T. P. (1977). Lyso-
    zyme-like activity in the hemolymph of *Biomphalaria glabrata*
    challenged with bacteria. *J. Invert. Pathol.*, 29, 170-174.
Cheng, T. C., Lie, K. J., Heyneman, D., and Richards, C. S. (1978).
    Elevation of aminopeptidase activity in *Biomphalaria glabrata*
    (Mollusca) parasitized by *Echinostoma lindoense* (Trematoda).
    J. Invert. Pathol., 31, 51-62.
Harris, K. R. (1975). The fine structure of encapsulation in
    Biomphalaria glabrata. *Ann. N.Y. Acad. Sci.*, 266, 446-464.
Harris, K. R. and Cheng, T. C. (1975a). The encapsulation process
    in *Biomphalaria glabrata* experimentally infected with the
    metastrongylid *Angiostrongylus cantonensis:* light microscopy.
    Intl. J. Parasitol., 5, 521-528.
Harris, K.R. and Cheng, T. C. (1975b). The encapsulation process
    in *Biomphalaria glabrata* experimentally infected with the
    metastrongylid *Angiostrongylus cantonensis:* enzyme histo-
    chemistry. *J. Invert. Pathol.*, 26, 367-374.

Krupa, P.L., Lewis, L. M., and Del Vecchio, P. (1977). *Schistosoma haematobium* in *Bulinus guernei:* electron microscopy of hemocyte-sporocyst interactions. *J. Invert. Pathol., 30,* 35-45.

Pan, C. T. (1963). Generalized and focal tissue responses in the snail *Australorbis glabratus* infected with *Schistosoma mansoni. Ann. N.Y. Acad. Sci., 113,* 475-485.

Pan, C. T. (1965). Studies on the host-parasite relationship between *Schistosoma mansoni* and the snail *Australorbis glabratus. Am. J. Trop. Med. Hyg., 14,* 931-976.

McDade, J. E. and Tripp, M. R. (1967). Notes on lysozyme in oyster mantle mucus. *J. Invert. Pathol., 9,* 581-582.

Newton, W. L. (1952). The comparative tissue reaction of two strains of *Australorbis glabratus* to infection with *Schistosoma mansoni. J. Parasitol., 38,* 362-366.

Newton, W. L. (1954). Tissue response to *Schistosoma mansoni* in second generation snails from a cross between two strains of *Australorbis glabratus. J. Parasitol., 40,* 352-355.

Newton, W. L. (1955). The establishment of a strain of *Australorbis glabratus* which combined albinism and high susceptibility to infection with *Schistosoma mansoni. J. Parasitol., 41,* 526-528.

Rachford, F. W. (1976). Host-parasite relationship of *Angiostrongylus cantonensis* in *Lymnaea palustris* II. Histopathology. *Exptl. Parasitol., 39,* 382-392.

Rodrick, G. E. and Cheng, T. C. (1974). Kinetic properties of lysozyme from *Crassostrea virginica* hemolymph. *J. Invert. Pathol., 24,* 41-48.

Sminia, T. (1972). Structure and function of blood and connective tissue cells of the fresh water pulmonate *Lymnaea stagnalis* studied by electron microscopy and enzyme histochemistry. *Zeit. Zellforsch. Mikroskop. Anat., 130,* 497-526.

Tripp, M. R. (1961). The fate of foreign materials experimentally introduced into the snail *Australorbis glabratus. J. Parasitol., 47,* 745-751.

Ubelaker, J. E., Cooper, H. B., and Allison, V. F. (1970). Possible defensive mechanism of *Hymenolepis diminuta* cysticercoids to hemocytes of the beetle *Tribolium confusum. J. Invert. Pathol., 16,* 310-312.

Yoshino, T. P. and Cheng, T. C. (1977). Aminopeptidase activity in hemolymph and body tissue of the pulmonate gastropod *Biomphalaria glabrata. J. Invert. Pathol., 30,* 76-79.

Effects of the Trematode Proctoeces maculatus on the Mussel Mytilus edulis

M. R. TRIPP

and

R. M. TURNER

School of Life and Health Sciences
University of Delaware
Newark, Delaware

I. INTRODUCTION

Members of the family Fellodistomatidae, like other digenetic trematodes, normally have a life cycle in which a fish hosts the adult worm, sporocysts mature in bivalve mollusks, and cercariae infect the fish. Members of the genus *Proctoeces*, however, have an unusual abbreviated life cycle (termed progenetic) in which both adult trematodes and sporocysts mature within the molluscan host. Adults of *Proctoeces maculatus* have been found in the kidney of the bivalve *Scrobicularia plana* (Freeman and Llewellyn, 1958) and in *Mytilus edulis* (Stunkard and Uzmann, 1959). When we discovered a population of *M. edulis* in Lewes, Delaware, infected with *P. maculatus*, we decided to study the seasonality of infection and the histopathology associated with the infection. This is a preliminary report of the first 10 months of this study.

II. MATERIALS AND METHODS

Mussels, *Mytilus edulis*, were collected from a single population attached to a floating dock at the University of Delaware Pollution Ecology Laboratory at Roosevelt Inlet, Lewes, Delaware. Approximately monthly samples were taken. Animals ranged in size between 3 and 7 cm in length but detailed measurements of individuals were not recorded. Mussels were examined within 24 hr of their return to the laboratory by removing one valve, looking for and removing adult flukes from the pericardial cavity, and fixing the whole animal in Bouin's solution in preparation for routine histological examination.

III. RESULTS

A. *SEASONAL INCIDENCE OF INFECTION*

1. *Sporocysts*

Larval trematodes were found in all samples examined over the period October, 1976, to July, 1977. The incidence of infection varied from 20 to 80% and averaged 36.3%. No attempt was made to estimate the severity of infection in individual mussels because of great variability but there was a tendency towards extreme parasite density in the fall months. No pronounced seasonal difference in relative numbers of mother and daughter sporocysts was detected.

2. *Adults*

Peak infection of the pericardial cavity of *M. edulis* by adult trematodes occurs in the fall, declines sharply with the onset of cold weather, and remains low through the summer. Evidence of the presence of earlier adult infection is manifest in masses of hemocytes in the pericardial cavity, in which trematode eggs or fragments of degenerating trematode tissue are embedded. Detection of mussels infected by adult trematodes is simple because these hemocyte masses are easily visible to the naked eye. It is not clear why the cellular response persists for so long after intact adults are no longer present.

3. Dual infection, adults and sporocysts

In a few cases both adult and larval worms were found in individual mussels. These double infections occurred only when adults were plentiful and apparently is a matter of chance.

B. PATHOLOGY

1. Sporocysts

There is no detectable hemocyte response by the molluscan host to the presence of healthy sporocysts (Figs. 1-3), a phenomenon usually reported when congenial mollusk-trematode infections are observed. Even with very heavy infections where considerable amounts of host tissue are occupied, no significant host cell response is detected. In instances where deteriorating sporocysts have been seen, invasion by host hemocytes is evident. Presumably this represents the clearing of foreign material by the host following death of the parasite. Although sporocysts were found most commonly and in greatest numbers among digestive diverticula, numerous other tissues, e.g., kidney, mantle, gonad, etc., also harbored sporocysts. Parasitic castration similar to that seen in other bivalves infected with trematodes, e.g., oysters with *Bucephalus*, was not observed and no evidence of decreased reproductive potential was obtained.

2. Adults

Young adult trematodes may be found in several tissues (mantle, foot, gonad, digestive diverticulum, kidney, connective tissue around adductor muscle) (Fig. 4) usually not eliciting any hemocytic response. On occasion, however, a pronounced hemocytic reaction is observed (Figs. 5, 9) possibly because the worm was dead or damaged.

The activity of mature trematodes in the pericardial cavity is dramatic. Adult worms attach to the outer wall of the cavity by the ventral sucker and can be seen to "massage" adjacent host tissues with the oral sucker. A dramatic hemocytic response usually (Fig. 6), but not always (Fig. 7), results. Host hemocytes are usually visible in the lumina of the digestive caeca (Figs. 6-9) in various stages of necrosis, suggesting active

Fig. 1.   Immature sporocyst containing germ balls (arrow) among
          tubules of digestive diverticulum of *Mytilus edulis*.
          H & E, 100X.

Fig. 2.   Mother sporocysts in kidney tissue of *Mytilus edulis*.
          H & E, 100X.

Fig. 3.   Daughter sporocysts containing cercariae (with suckers,
          arrows) in digestive diverticulum of *Mytilus edulis*.
          H & E, 100X.

Fig. 4.   Young adult *Proctoeces* in mature testis of *Mytilus
          edulis*. H & E, 100X.

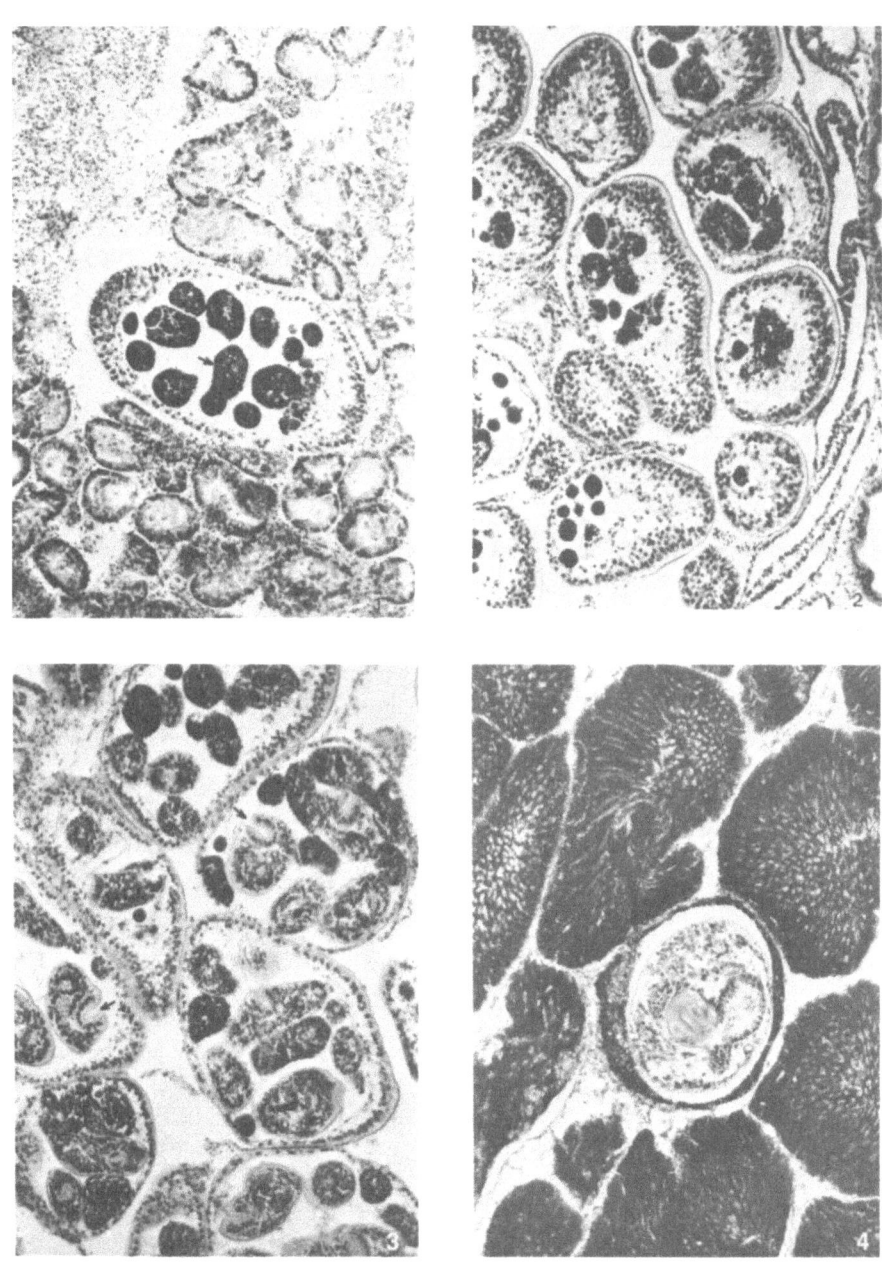

Fig. 5. Young adult *Proctoeces* in mature testis of *Mytilus edulis;* note host hemocyte (HC) reaction (compare with Fig. 4). H & E, 100X.

Fig. 6. Adult *Proctoeces* in pericardial cavity with massive host hemocyte (HC) response. Oral sucker (OS), esophagus (E), digestive caecum (DC) and ventral sucker (VC) are readily identified landmarks. Note hemocytes in digestive caecum and in cavity of ventral sucker (arrows). H & E, 100X.

Fig. 7. Sagittal section of adult *Protoeces* in pericardial cavity with minimal hemocyte response by host *Mytilus edulis*. Note hemocytes in digestive caecum (DC). H & E, 100X.

Fig. 8. Two adult *Proctoeces* in pericardial cavity of *Mytilus edulis* embedded in mass of host hemocytes. Note hemocytes in digestive caeca. H & E, 100X.

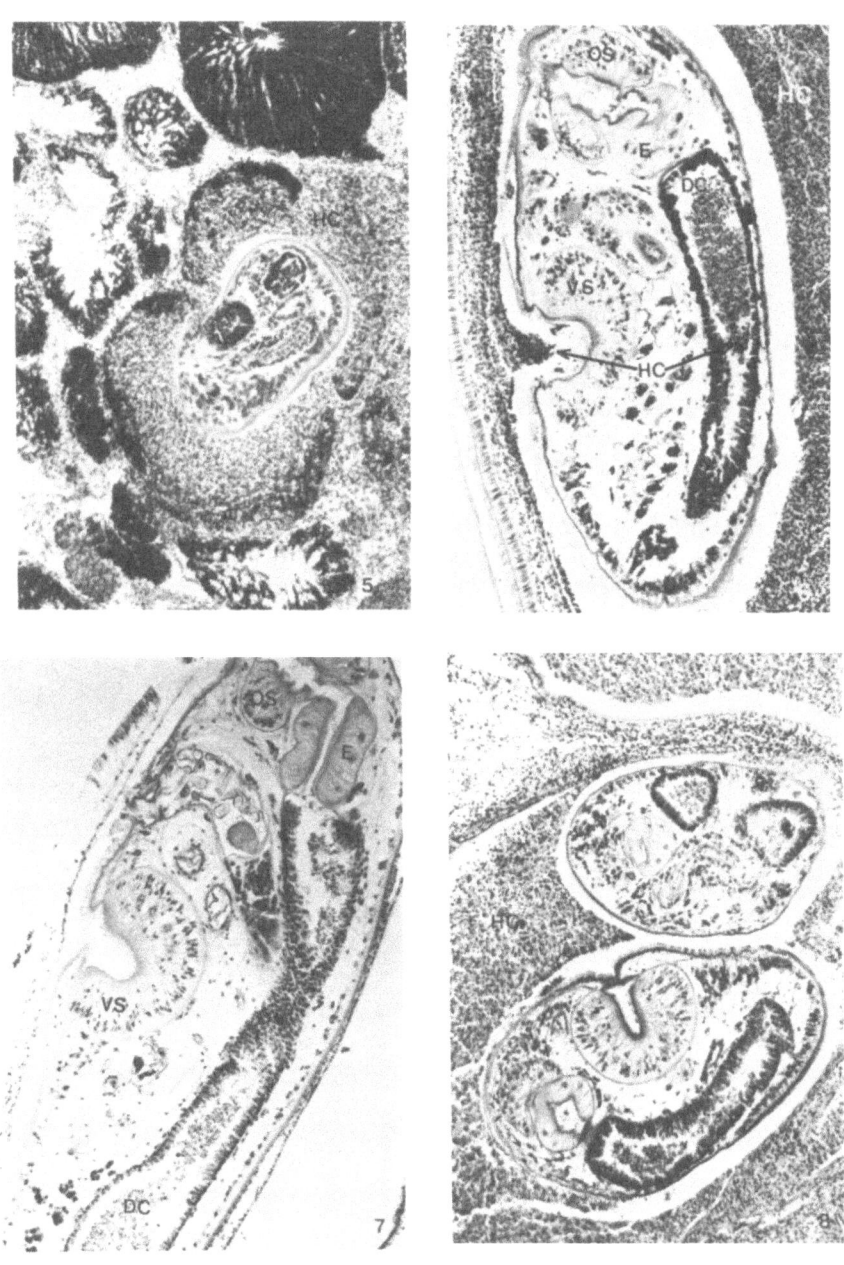

Fig. 9.   Adult *Proctoeces* in foot of *Mytilus edulis* surrounded
         by host hemocytes. H & E, 100X.

Fig. 10. Adult *Proctoeces* in foot of *Mytilus edulis* surrounded
         by hemocytes and undergoing degeneration. Digestive
         caeca (DC) and ventral sucker (VS) are still recognizable
         but other tissues are degenerating. H & E, 100X.

Fig. 11. Adult *Proctoeces* in pericardial cavity of *Mytilus
         edulis* undergoing degeneration. Oral sucker (OS) and
         ventral sucker (VS) with hemocytes in cavities and
         eggs (Eg) are distinguishable; other tissues necrotic.
         H & E, 100X.

Fig. 12. Degeneration of adult *Proctoeces* in *Mytilus edulis*
         pericardium almost complete with eggs (Eg) still visible
         and hemocytes (HC) invading and replacing rest of
         trematode tissues. H & E, 100X.

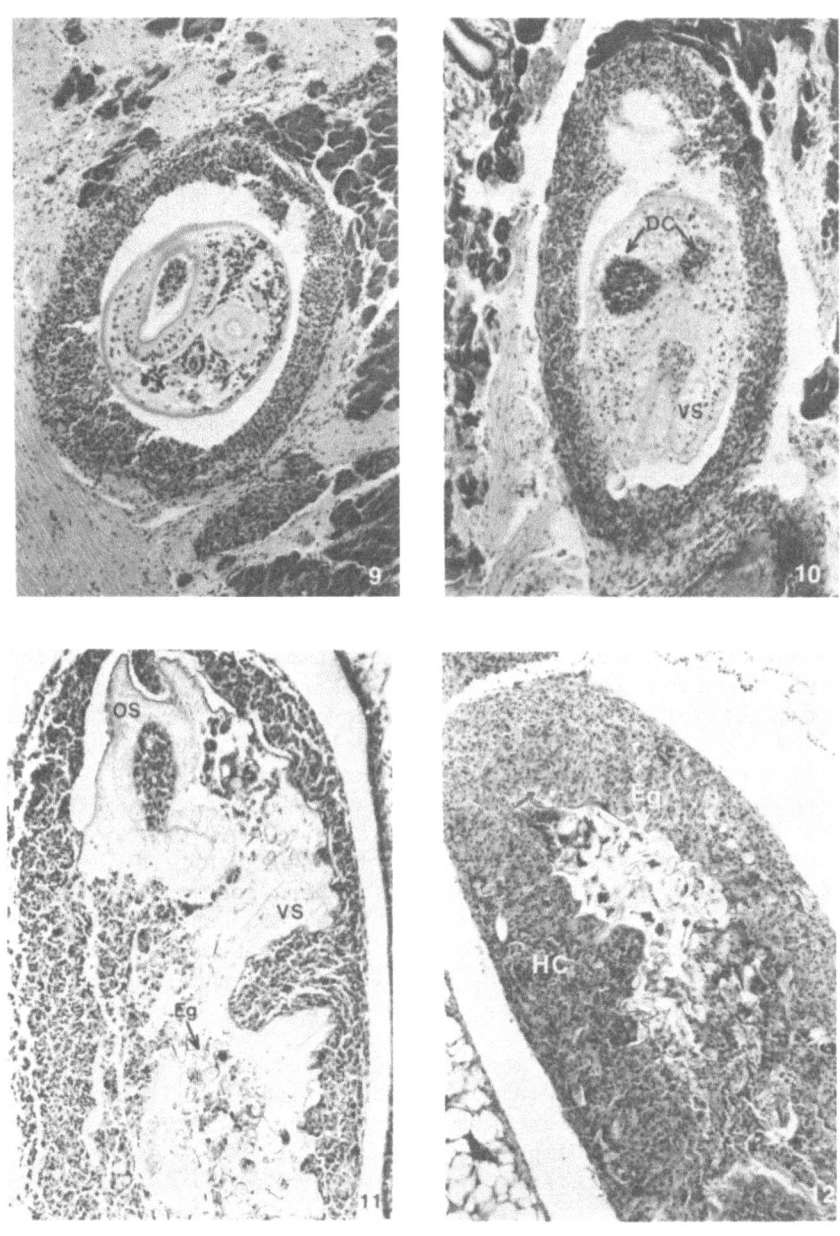

ingestion and digestion by the trematode. Most often one to three
worms are seen in the pericardium but on several occasions ten to
twelve have been found, and in one instance twenty-five. Healthy
adult worms in tissues other than the pericardium may be surrounded by hemocytes also (Fig. 9), but it is not know if that is the
result of feeding or simply a response to the presence of foreign
material.

Adult trematodes enveloped by hemocytes can be seen to be
undergoing degeneration (Fig. 10). There is a progressive invasion and degradation of trematode tissue by hemocytes (Fig. 11)
until only the vague outline of the worm and residual eggs are
detectable (Fig. 12). These masses of hemocytes with occasional
embedded eggs persist in the pericardium in the absence of any
other evidence of infection, presumably even months after active
infection.

VI. DISCUSSION

The seasonal incidence of infection of *Mytilus edulis* with
Proctoeces maculatus in Delaware is similar to that at Shark
River, New Jersey, as reported by Lang and Dennis (1976). They
reported sporocysts in 20-30% of the dock mussels examined from
April through November, with a decreased incidence of infection
(5-12%) from December through March. With the exception of our
November sample, the incidence of sporocysts in the mussels we
sampled was 20-40% throughout the year. There was no obvious
correlation between season and stage of development of sporocysts
as reported by Uzmann (1953). At most times mussels contained a
mixture of mother and daughter sporocysts, although individual
hosts might show a predominance of one or the other. Conceivably
this is due to the generally warmer Delaware waters; Uzmann
(1953) studied mussels from Long Island Sound.

Sporocysts of *P. maculatus* have been described as occurring
primarily in the venous sinuses, particularly of the mantle.
Uzmann (1953) postulated that this reduced the efficiency of
circulation and distribution of nutrients essential for gonadal
development and led to parasitic castration. Dupouy and Martinez
(1973) have challenged this concept and suggested instead that
once gonadal development was initiated the parasite had little
effect on sexual maturation. We find that sporocysts are distributed among many tissues, predominantly in digestive diverticula, but also in gonad, mantle, kidney, or almost any other
tissue except heart. We also failed to find evidence of
parasitic castration. Even in the presence of heavy infections
reproduction seemed unimpaired.

Perhaps the most striking finding we have made is the vigorous
activity of the adult worms in the pericardial cavity and the
mollusks marked response to them. The adult trematodes seem to
"graze" on the surface of the pericardial tissues, ingesting cells

Table 1. Percent *Mytilus edulis* infected with *Proctoeces maculatus*

| Date | N | % Mussels with | | | |
		Sporo.	Adults	Adults + Eggs	Adult + Sporo.
1976					
Oct.	100	20	45	67	7
Nov.	33	80	35	50	30
Dec.	25	33	32	100	0
1977					
Feb.	12	33	8	58	0
Mar.	25	20	4	72	0
Apr.	41	46	0	14	0
June	18	22	6	39	0
July	45	36	4	8	2
Total	299				
Aver.	37.4	36.3	16.8	51.0	4.9

and mucus, and eliciting a strong hemocyte response. Hemocytes may attach to the surface of the worm but apparently do little to hinder its movement. Eggs are released by the worm and become embedded in hemocyte masses that persist indefinitely, even in the absence of adult worms. During the winter months, the worms apparently are less able to cope with the hemocytes, become enveloped by them and die. Whether death is due to hemocyte action, cold weather (or both), or to some other cause, we do not know. However, dead or dying worms are invaded by hemocytes, the carcass is broken down and the infection resolved. Thus with the coming of warm weather and the shedding of new cercariae the limited space in the mussel pericardium becomes available for maturing adult trematodes.

REFERENCES

Dupouy, J. et J-C. Martinez. (1973). Action de *Proctoeces maculatus* (Looss, 1901)(Trematoda, Fellodistomatidae) sur le développement des gonades chez *Mytilus galloprovincialis* Lmk. *C.R. Acad. Sc. Paris, Série D t.* 277, 1889-1892.
Freeman, R. F. H. and Llewellyn, J. (1958). An adult digenetic trematode from an invertebrate host: *Proctoeces subtenuis* (Linton) from the lamellibranch *Scrobicularia plana* (Da Costa). *J. Mar. Biol. Ass. U.K.,* 37, 435-457.
Lang, W. H. and Dennis, E. A. (1976). Morphology and seasonal incidence of infection of *Proctoeces maculatus* (Looss, 1901) Odhner, 1911 (Trematoda) in *Mytilus edulis* L. *Ophelia* 15, 65-75.
Stunkard, H. W. and Uzmann, J. R. (1959). The life-cycle of the digenetic trematode, *Proctoeces maculatus* (Looss, 1901) Odhner, 1911) [Syn. *P. subtenuis* (Linton, 1907) Hanson, 1950], and description of *Cercaria adranocerca* n. sp. *Biol. Bull.,* 116, 184-193.
Uzmann, J. R. (1953). *Cercaria milfordensis* nov. sp., a micro-cercous trematode larva from the marine bivalve, *Mytilus edulis* L. with special reference to its effect on the host. *J. Parasitol.,* 39, 445-451.

The Role of Hemocytes in Melanotic Tumor Formation

T. M. RIZKI

and

ROSE M. RIZKI

Division of Biological Sciences
University of Michigan
Ann Arbor, Michigan

I. INTRODUCTION

The diseases of insects may be grouped in two main categories:
those in which the pathological syndrome is generated by the
presence of an infective agent, and those in which a genetic
defect underlies the pathological condition. Hemocytes are involved in combating deleterious effects in both cases. In the
case of infection we can be certain that the hemocytes are
reacting to stimuli emitted by the presence of the infective
agent. Cell-mediated defense responses of insects to foreign
bodies in the hemocoel include phagocytosis and nodule formation
by the hemocytes (Salt, 1970). Similar types of behavior are
exhibited by hemocytes in some genetic diseases (Oftedal, 1953).
However, in the latter cases, analysis of the pathological syndrome is complicated by the difficulty in determining whether
the genetic lesion directly affects the hemocytes or the hemocytes are reacting to gene-conditioned disorders elsewhere in
the body.

II. MELANOTIC TUMORS

Melanotic tumor formation in *Drosophila melanogaster* is an
example of inherited disease in which the cell-mediated defense
system of the body plays a major role. We have been studying
two mutant genes that cause melanotic tumors in adipose tissue:
tu-W (2 – 63.8) and *tu-ts*^{Sz} (1 – 34.3). Since these are non-
allelic genes, the genetic basis of tumor formation in the strains
is different. Our studies indicate that the hemocytes undergo
similar changes in the two mutants to produce the final phenotype, melanotic tumor, but the sequence of morphological events
at the tumor-forming site in the mutants differs.

Melanotic tumors in *tu-W* and *tu-ts*^{Sz} larvae appear in the late
third instar in the posterior region of the fat body. The melanotic phenotype is temperature conditioned in the *tu-ts*^{Sz} mutant:
tumors appear in larvae grown at 26°C but not in larvae reared
at 18°C. In both mutants the melanotic tumors are retained in
the pupal stage and adult life. They are not lethal to the
individuals bearing them although fecundity and longevity in the
tu-W mutant are reduced (Wilson *et al.*, 1955). Interrelationships
of the known effects of the *tu-W* gene on development include
hormonal centers in the ring gland (Rizki, 1960) as well as the
cellular components directly associated with tumor formation
(Fig. 1). A catalog of the developmental effects of the *tu-ts*^{Sz}
gene has not been completed as yet.

Microscopical comparisons of adipose cells and hemocytes from
tumor strains have been made with a tumor-free wild type strain
of *D. melanogaster (Ore-R)*. Hemolymph samples from third instar
Ore-R larvae examined with the phase microscope or in stained

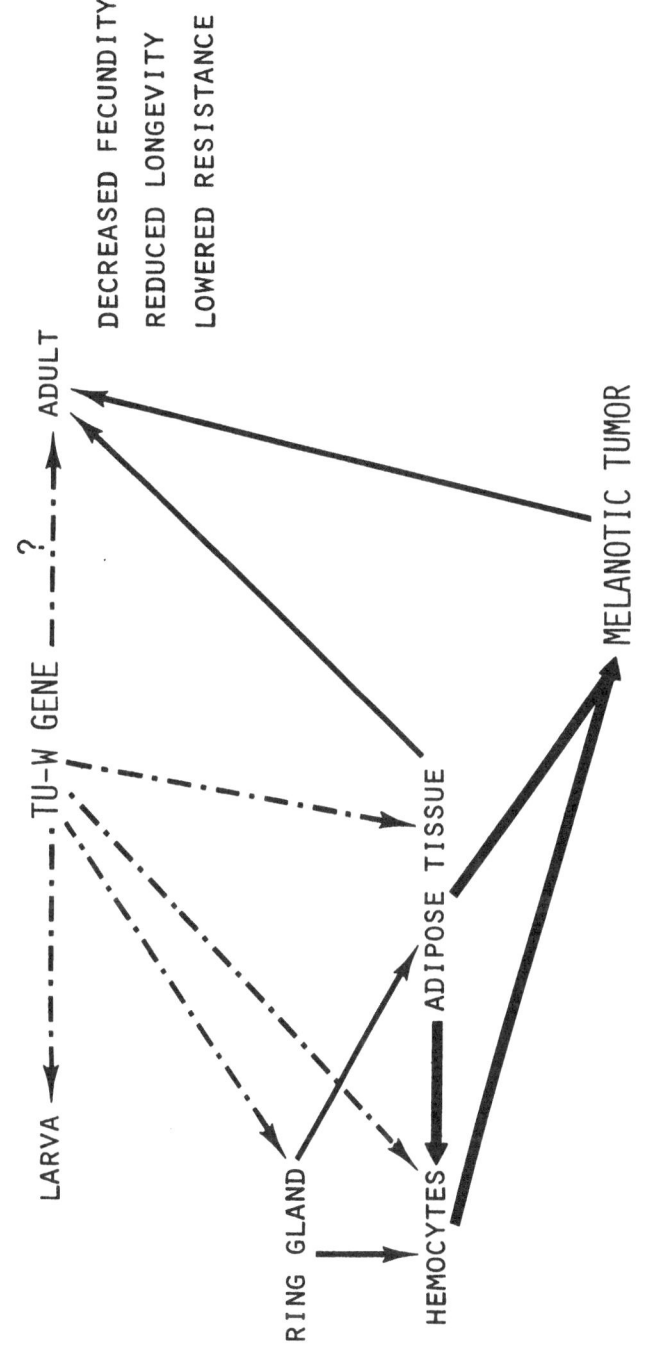

Fig. 1. The *tu-W* syndrome. The heavy solid lines indicate the interaction between adipose tissue and hemocytes to produce melanotic tumors. Thin solid lines represent epigenetic relationships resulting from the metabolism of the ring gland and the adipose tissue. Since the primary site of *tu-W* gene action is unknown, possible spheres of gene influence are indicated by broken lines.

smears contain plasmatocytes and crystal cells (Rizki, 1957).
The latter, which constitute 5 - 10% of the hemocyte population, have large cytoplasmic inclusions (Rizki et al., 1976).
We have postulated that these cells carry melanin precursors that
may be used in melanization of tumors (Rizki and Rizki, 1959),
but their direct participation in tumor formation has not been
demonstrated. The plasmatocytes and their derivatives are the
hemocyte components of melanotic tumors. Plasmatocytes are
spherical cells that are transformed into podocytes and a
flattened variant, the lamellocyte (Rizki, 1957b, 1962). Lamellocytes are abundant in hemolymph samples from the $Ore-R$ strain
following pupariation but are infrequent in larval samples.
In $tu-W$ and $tu-ts^{Sz}$ larval hemolymph many lamellocytes are found
throughout the third instar. They can be seen in freshly drawn
hemolymph samples examined with the phase microscope as well as
in larvae fixed for examination with the scanning electron microscope. In the latter method cells that are free in the hemolymph
are lost when the body cavity is opened during the fixation
process so only those hemocytes adhering to tissue and body wall
surfaces or entrapped within spaces between internal organs remain for examination in the scanning electron microscope. Lamellocytes are common in third instar $tu-W$ and $tu-ts^{Sz}$ larvae fixed
in this manner but not in $Ore-R$ larvae.

III. SITE OF TUMOR FORMATION

Since the site of tumor formation is constant in $tu-W$ and $tu-$
ts^{Sz} larvae, the prospective tumor regions can be examined to
seek pretumorous changes. In $tu-W$ larvae the first visible
change at the prospective tumor site is a rounding up of the
caudal fat body cells, apparently due to a loss of intercellular
adhesion. Small globules begin to ooze from the intercellular
spaces and the basement membrane begins to break down or dissolve.
Hemocytes appear on the surface of the fat body specifically
at these affected points. Electron microscopy of thin sections
of the tumor-forming site confirms the loss of basement membrane
and cytoplasmic contents of the adipose cells, and shows that
the hemocytes attracted to this area engulf these materials
(Figs. 2,6). Hemocytes continue to aggregate on the surfaces of
the fat body cells and a layering of the lamellocytes persists
until the entire affected region of the caudal fat body is firmly
encased (Fig. 3). Melanization of the encapsulated cellular
masses is finalized during late larval life.
Melanotic tumors in $tu-ts^{Sz}$ are also formed by encapsulation of
adipose cells in larvae raised at 26°C (Figs. 4, 5, 7). In this
mutant intracellular degenerative processes in the form of
membranous whorls or myelin forms can be observed before any
changes of the tissue or cell surfaces are recognizable by scanning electron microscopy. Furthermore, some of the adipose cells

in the prospective tumor site are distinctly deficient in lipid droplets and rich in ER whereas other cells look normal with well formed lipid droplets. Tumors do not form in tu-ts^{Sz} larvae at the nonpermissive temperature of 18°C, and massive intracellular degenerative changes are not apparent during the early third instar at this temperature. When the fat body cells of these larvae are compared with Ore-R cells at 18°C, the size and frequency of lipid droplets in both strains are similar. Additional quantitative studies of the lipid and glycogen contents of tu-ts^{Sz} cells are contemplated to determine whether total normalization is achieved at 18°C. The answer to this question is important for analyzing the cause and effect relationships in melanotic tumor formation since the hemolymph of tu-ts^{Sz} larvae at 18°C contains many lamellocytes.

IV. MORPHOLOGY OF TUMOR FORMATION

The sequence of morphological steps in melanotic tumor formation is similar in tu-W and tu-ts^{Sz} larvae. Lamellocytes actively enclose regions of the fat body that show degenerative changes, and we can clearly perceive this phenomenon as a protective device to rid the body of aberrant cells. The specificity in the aggregation of lamellocytes to the afflicted tissues suggests hemocyte recognition of healthy vs unhealthy cells. Presumably the stimuli responsible for this interaction are chemical and are limited to the immediate vicinity of the aberrant cells. Phagocytic engulfment of particulate materials by hemocytes resting on caudal fat body cells occurs in both mutants suggesting that the unscheduled exit of intracellular materials from adipose cells is a factor initiating hemocyte activity in the body.

In both mutants lamellocytes appear in the hemolymph much earlier than the time at which encapsulation of adipose tissues begins. During this interval the lamellocytes are passively circulating in the hemocoel. The hemolymph enters the caudal ostia of the heart, flows out from the anterior opening of the dorsal aorta, and returns to the caudal region of the body via the lateral sinuses of the hemocoel (Rizki, 1978). We can surmise that the circulating lamellocytes as well as the other hemocytes make random contacts with the presumptive tumor site, yet adhesion of lamellocytes does not occur until changes in the topology of the caudal fat body become apparent. Surface features of the adipose tissue must be an important factor in initiating the process of lamellocyte adhesion to form capsules. What the factors are that stimulate the transformation of plasmatocytes and podocytes to lamellocytes, however, are not clear.

Fig. 2.   Early stage of tumor formation in tu-W. Note the
          accumulation of intercellular material and the rounding
          up of four adipose cells. The small hemocyte is adhering
          to the membranous material and the surface of the adipose
          cell in the lower right corner. Compare with Figure 6.

Fig. 3.   Late stage in the encapsulation process in tu-W. Note
          the large lamellocyte covering the adipose tissue
          surface at the left side of the frame and the numerous
          hemocytes in various stages of flattening. The meandering
          tubular structure is a trachea. A portion of a fully
          encapsulated tumor surface is shown in the upper right
          quarter.

Fig. 4.   Encapsulation of tu-ts^{Sz} adipose cells by hemocytes.
          Numerous lamellocytes as well as podocytes can be seen
          on the fat body surface. There seems to be no order to
          the layering process; flat cells may layer over or under
          the spherical form cells.

Fig. 5.   A fully formed melanized tumor removed from tu-ts^{Sz} larva.
          Note the compact nature of the tumor due to the wrapping
          by lamellocytes. Compare with Fig. 7.

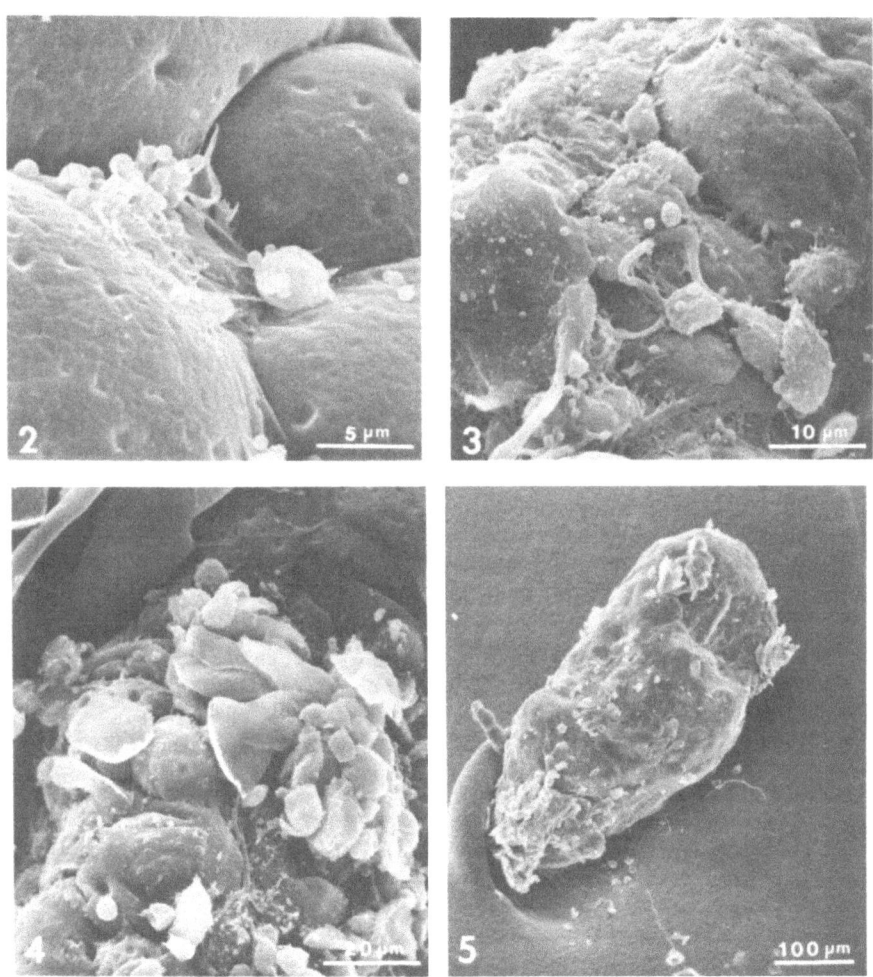

Fig. 6.   A tumor-forming site in *tu-W* showing phagocytosis of
          basement membrane material. The podocyte (p) is resting
          at the surface of the intercellular space between two
          fat body cells (f). Note the presence of basement
          membrane (bm) on the surface of the upper fat body cell
          and its absence from the surface of the lower fat body
          cell (arrowheads). The vesicles (v) correspond to the
          globules and membranous elements at the intercellular
          space in Figure 2. Formaldehyde-osmic fixation with
          ruthenium red/uranyl acetate-lead citrate stain.
          Scale = 1 μm.

Fig. 7.   Section showing part of an encapsulated melanized tumor
          in *tu-ts^{Sz}*. Note the layering of lamellocytes (arrowheads)
          at the surface of the degenerating fat body cells (f).
          The outermost lamellocyte is sectioned through the
          nucleus (n); note also the nature of the cytoplasmic
          inclusions in this cell. Formaldehyde-osmic fixation/
          uranyl acetate-lead citrate stain.   Scale = 1 μm.

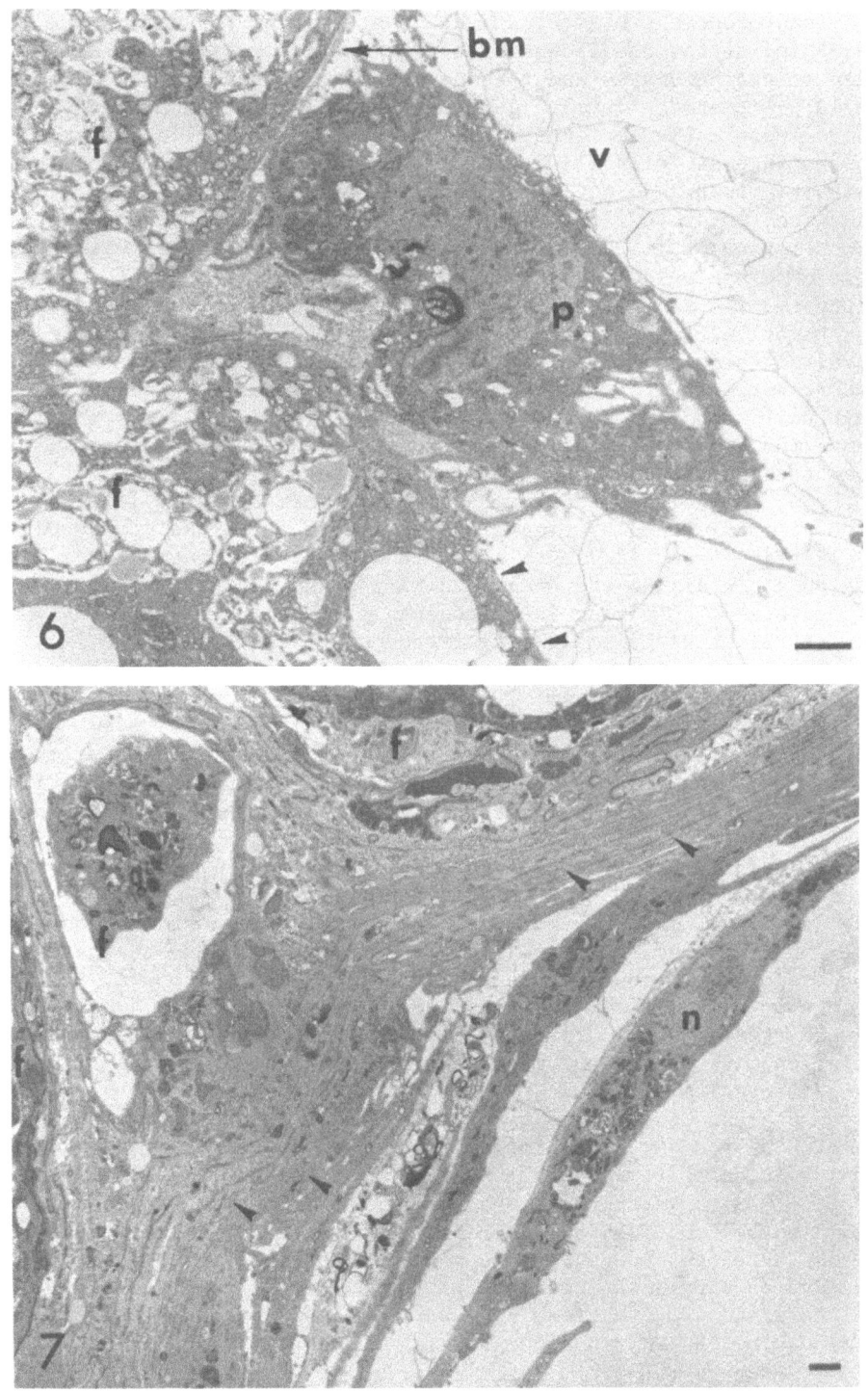

We can propose two alternatives in attempting to unravel the
cause and effect relationships of the morphological transformation of the hemocytes and the degenerative changes in the adipose
cells: either the primary genetic lesion affects both hemocytes
and adipose cells, or genetically determined metabolic defects
in the adipose cells elicit lamellocyte formation. Even though
melanotic tumors do not develop in tu-ts^{Sz} larvae at 18°C,
lamellocytes are present in high frequency. Therefore, melanotic
tumor formation is not obligatory in the presence of lamellocytes.
This observation does not preclude the second alternative,
however. The factors initiating lamellocyte formation may differ
quantitatively and/or qualitatively from the factors affecting
lamellocyte aggregation, and in the latter instance, both may
emanate from the fat body cells. It is reasonable to suggest
that early biochemical changes in the adipose cells alter the
hemolymph, and the circulating hemocytes sense these signals
that precede the massive degenerative changes.

The morphological transformation of plasmatocytes to lamellocytes is influenced by ring gland hormones. This was demonstrated
by ligating Ore-R larvae to separate the anterior segments of
the body containing the brain and ring gland from the posterior
segments, and comparing the frequency of lamellocytes in the
latter region with lamellocyte frequency of control larvae
ligated to include the brain and ring gland in the posterior
region of the body (Rizki, 1962). Lamellocyte frequency was
increased when the hormone centers were excluded, and this
effect was blocked by implanting ring glands into the posterior
portion of the body. Hormones from the ring gland also affect
the incidence of tumors in melanotic tumor strains (Burdette,
1954), and this relationship has been examined in the tu-W
mutant (Rizki, 1960). More recently Madhavan (1972) succeeded
in inducing melanotic tumors in nontumor strains by topical
application of juvenile hormone. Since the ring gland hormones
are important for the initiation of metamorphosis and the onset
of degenerative processes in the larval tissues, hemocyte
changes resulting from hormonal modifications may be a conse-
quence of degenerative processes induced in other tissues. Direct
relationships of these factors can best be evaluated by in $vitro$
experimentation.

The process of encapsulation and melanization to form melanotic
tumors in $Drosophila$ is similar to the hemocytic reaction elicited
by infection of $Drosophila$ larvae with parasites (Walker, 1959;
Nappi, 1973). Furthermore, bacterial infections in $Drosophila$
may also be contained by a similar type mechanism (Rizki, 1969).
It is therefore understandable that the possibility of an infective or viruslike agent as the causal factor in hereditary
melanotic tumor formation in $Drosophila$ has often been raised
(for review, see Harshbarger and Taylor, 1968). Conclusive

demonstration of such an etiological agent by injection and implantation techniques is complicated since a variety of implanted foreign materials as well as tissue injury will induce melanotic masses in insects. In an electron microscopic examination of the cells of tumor strains Perotti and Bairati (1968) did not find viruslike particles that were specific to the tumorous cells, and ultrastructural comparisons of tu-W and tu-ts^{Sz} with the nontumorous strain, Ore-R, have given no indication of an infective agent in the affected adipose cells. Clear-cut evidence of a hereditary metabolic disorder is evident among the caudal adipose cells of tu-ts^{Sz} larvae strengthening the argument that tumorigenesis in these cases results from a defense response of the hemocytes against diseased cells.

ACKNOWLEDGMENT

This investigation was supported by Grant No. CA-16619 awarded by the National Cancer Institute, DHEW, and in part by NIH Biochemical Sciences Grant No. RR-07050.

REFERENCES

Burdette, W. J. (1954). Effect of ligation of *Drosophila* larvae on tumor incidence. *Cancer Res.*, 14, 780-782.

Harshbarger, J. C. and Taylor, R. L. (1968). Neoplasms of insects. *Ann. Rev. Entomol.*, 13, 159-190.

Madhavan, K. (1972). Induction of melanotic pseudotumors in *Drosophila melanogaster* by juvenile hormone. *Wilhelm Roux' Archiv.*, 169, 345-349.

Nappi, A. J. (1973). Hemocytic changes associated with the encapsulation and melanization of some insect parasites. *Exp. Parasitol.*, 33, 285-302.

Oftedal, P. (1953). The histogenesis of a new tumor in *Drosophila melanogaster* and a comparison with tumors of five other stocks. *Zeit. Induk. Abstammungs-. Vererbung.*, 85, 408-422.

Perotti, M. E. and Bairati, A. (1968). Ultrastructure of the melanotic masses in two tumorous strains of *Drosophila melanogaster* (tu B_3 and Freckled). *J. Invert. Pathol.*, 10, 122-138.

Rizki, T. M. (1957). Alterations in the hemocyte population of *Drosophila melanogaster*. *J. Morph.*, 100, 437-458.

Rizki, T. M. (1957b). Tumor formation in relation to metamorphosis in *Drosophila melanogaster*. *J. Morph.*, 100, 459-472.

Rizki, T. M. (1960). Melanotic tumor formation in *Drosophila*. *J. Morph.*, 106, 147-158.

Rizki, T. M. (1962). Experimental analysis of hemocyte morphology in insects. *Amer. Zool.*, 2, 247-256.

Rizki, T. M. (1969). Hemocyte encapsulation of *Streptococci* in *Drosophila*. *J. Invert. Pathol.*, 12, 339-343.

Rizki, T. M. (1978). Circulatory System. *In:* "The Genetics and Biology of *Drosophila*", Vol. 2 (M. Ashburner and T. R. F. Wright, eds.). Academic Press, London.

Rizki, T. M. and Rizki, R. M. (1959). Functional significance of the crystal cells in the larva of *Drosophila melanogaster*. *J. Biophys. Biochem. Cytol.*, 5, 235-240.

Rizki, T. M., Rizki, R. M., Allard, L. F., and Bigelow, W. C. (1976). Micromanipulation of tissues and cells of the *Drosophila* larva in the SEM. *Scanning Electron Microscopy*, *IITRI* 2, 611-618.

Salt, G. (1970). "The Cellular Defense Reactions of Insects." Cambridge University Press.

Walker, I. (1959). Die Abwehrreaktion des Wirtes *Drosophila melanogaster* gegen die Zoophage Cynipidae *Pseudeucoila bochei* Weld. *Rev. Suisse Zool.*, 66, 569-632.

Wilson, L. P., King, R. C., and Lowry, J. L. (1955). Studies on the *tu-W* strain of *Drosophila melanogaster*. I. Phenotypic and genotypic characterization. *Growth*, 19, 215-244.

A Study of Granuloma Formation by Molluscan Cells[1]

THOMAS C. CHENG

Institute for Pathobiology
Center for Health Sciences
Lehigh University
Bethlehem, Pennsylvania

[1]This research was supported by Grant AI 12355-03 from the National Institute of Allergy and Infectious Diseases, U.S. Public Health Service.

I. INTRODUCTION

The development, nature, and functional significance of infectious granulomatoses in mammals, particularly humans, have held
a position of central interest in immunopathology for decades.
First recognized in patients who had died from tuberculosis,
these inflammatory cellular aggregates, referred to as "tubercules,"
were identified as being comprised of large phagocytic cells or
histiocytes. Furthermore, a prominent feature of each granuloma
is the formation of multinucleated giant cells, containing a
peripheral zone of lymphocytes, with or without plasma cells.

Although the development of granuloma is usually associated
with invasion by infectious agents, e.g., tuberculosis, histo-
plasmosis, coccidiomycosis, actinomycosis, blastomycosis,
schistosomiasis, capillariasis, etc., this need not be always the
case. It is known, for example, that noninfectious materials,
such as silica, can serve as the focus for the development of
granuloma.

Generally, granulomatous inflammation is an undesirable aspect
of cellular response since, as in the case of helminthic infections, tuberculosis, and syphilis, there is commonly extensive
destruction of tissues involving necrosis and fibrosis.

Granuloma development in vertebrates may or may not have an
immunologic basis. As an example of the first, Warren *et al.*
(1967) have demonstrated that in experimental murine schistosomiasis,
intact immunologic response is a prerequisite to the formation of
granulomas surrounding eggs lodged in tissues. Furthermore, the
administration of a variety of immunosuppressants totally or
partially inhibits granuloma formation. These evidences have led
to the conclusion that there is an immunologic basis. Also,
Warren *et al.* (1967) have demonstrated that presensitization augments
the development of granuloma. The accelerated augmented reaction is
not only specific but also is transferable with cells but not with
serum. It has since been demonstrated that the immunological basis
is of the delayed cell-mediated hypersensitivity type (Domingo
et al., 1967; Domingo and Warren, 1948a, b; Perrotto and Warren,
1969; Warren, 1969).

Subsequent studies by Solomon and his colleagues, involving use
of eggs of the nematode *Capillaria hepatica* in mice as the model,
have revealed additional information pertaining to parasitic
granuloma formation in mammals. Specifically, Solomon and Soulsby
(1973) have reported that previous intraperitoneal sensitization
of mice with eggs resulted in earlier and enhanced granuloma
development around eggs in the liver. This, of course, is further
evidence for an immunologic basis for granuloma development.
Furthermore, these investigators were able to demonstrate that
specificity was involved since there was a lack of enhanced
reaction to presensitization with eggs of a closely related
nematode, *Trichuris muris*.

Raybourne *et al.* (1974) have demonstrated that both agglutinating and homocytotropic antibodies, as well as delayed dermal reactivity, occur in mice with primary and secondary granulomas formed in reaction to *C. hepatica* eggs. Subsequently, Raybourne and Solomon (1975) were able to demonstrate that secondary granuloma formation produced an antibody response characterized by the initial production of IgM followed by IgG_1 and IgG_2 during the latter phase of the test period. On the other hand, the sera of mice with primary granulomas portrayed a more varied antibody response, with all three types of immunoglobulins being present during the entire test period. Also, 48 hr passive cutaneous anaphylaxis tests demonstrated the presence of reagin activity in sera from granulomatous mice. Thus, additional proof has been provided for an immunologic basis for granuloma development in mammals in response to helminth parasites.

The evidence that the second category of granuloma development in mammals is nonimmunogenic stems from studies involving the insertion of inert substances, such as divinyl-benzene-copolymer beads, into hosts (Warren and Kellermeyer, 1968; Warren and Domingo, 1970). The fact that encapsulating granulomas develop around such beads indicates that the cellular response is not directed toward an immunogen but represents a nonspecific inflammatory response to inert foreign material.

The brief review presented above points out that two categories of granulomas occur in mammals. The formation of one type is apparently mediated by immunoglobulins, although the exact mechanism remains unclear. The second type is apparently nonspecific, nor is it initiated by immunogens. Since molluscs, and all invertebrates for that matter, do not synthesize immunoglobulins, it would appear that the mechanism of granuloma development in these "lower" animals is more akin to the second type of mammalian granuloma formation, and detailed probes into the underlying mechanisms as they occur in molluscs would contribute to our understanding of at least one type of mammalian granuloma. Also, because of the absence of immunoglobulins in molluscs, the mechanism underlying nonantibody-mediated granuloma formation, which is considered to be of a more primitive type, can be studied by employing a variety of foreign bodies, both immunogenic and inert, without having to decipher if the granuloma had been formed as a result of immunoglobulin mediation.

Since it is well known that granuloma formation accompanies the destruction of larval schistosomes that penetrate incompatible snail hosts, I have employed this system as the model to study granuloma development.

II. MATERIALS AND METHODS

The specimens of the pulmonate gastropod *Biomphalaria glabrata* susceptible to Puerto Rican *Schistosoma mansoni* used in this study

were randomly selected from a stock of the so-called NIH albino
strain (Newton, 1955) which has been maintained at this institute
for over 7 years. For the sake of convenience, this strain is
referred to herein as the LU_{as} strain. Specimens of the Brazilian
strain employed, designated as LU_{bp}, represented the progeny of
snails collected at Belo Horizonte, Brazil, in 1973 by Dr. M. A.
Stirewalt. This strain is partially refractile to the strain of
S. mansoni which we used. The totally incompatible albino snails
used originated from a stock supplied by Dr. C. S. Richards of the
National Institute of Allergy and Infectious Diseases, Bethesda,
Maryland, and designated by him (Richards, 1975) as strain 10-R2.
All snails were maintained as previously described (Harris and
Cheng, 1975).

Eggs of a strain of Puerto Rican *S. mansoni* were harvested
from the livers of white mice according to the procedure of
Lim and Heyneman (1972). The infections in mice were 5 to 7 weeks
old at the time the eggs were isolated. Exposure of snails to
miracidia was initiated within 30 min of hatching.

A. ACID PHOSPHATASE IN ISOLATED CELLS

Since Yoshino and Cheng (1976) have demonstrated that in the hard
clam, *Mercenaria mercenaria,* the so-called cytoplasmic granules of
granulocytes are actually lysosomes and Jeong and Heyneman (1976)
have found rich concentrations of acid phosphatase in granulocytes
of *B. glabrata,* it was decided to employ this lysosomal enzyme as
a marker to trace the role of granulocytes during the encapsulation
of incompatible larval schistosomes. However, before this strategy
could be carried out, it had to be determined if this lysosomal
marker enzyme (deDuve *et al.*, 1955; deDuve, 1969; Novikoff, 1963)
is only present in granulocytes. Consequently, cytochemical tests
were carried out on both isolated granulocytes and hyalinocytes.

Hemocytes employed in our cytochemical studies were collected
from the three strains of *B. glabrata* by the method described
earlier (Cheng and Yoshino, 1976), which involved inserting a glass
capillary into the heart. The hemolymph samples were subsequently
expelled onto coverglasses, which were placed in a moist chamber,
i.e., a covered Petri dish the bottom of which was lined with moist
filter paper, maintained at 22°C for 15 min. This procedure permitted the cells to adhere and form a monolayer. The cells were
then fixed for 20 min in 0.5% glutaraldehyde in Sörensen's buffer,
pH 6.5, followed by four washings in water.

Gomori's (1950) lead nitrate method for the demonstration of
acid phosphatase was employed. This involved the utilization of
sodium β-glycerophosphate as the substrate. Identical preparations
which were incubated in a medium devoid of this substrate or in a
complete incubation medium to which had been added 0.01M of sodium
fluoride as an inhibitor served as the controls.

B.   ACID PHOSPHATASE IN CELLS PARTICIPATING IN ENCAPSULATION

In order to trace the involvement of granulocytes in the encapsulation of *S. mansoni* mother sporocysts, 20 snails, each measuring
3 to 5 mm in shell diameter, were randomly selected from stocks of
each of the three strains and exposed to miracidia in the following
manner.   Each snail was exposed to 30 miracidia in a 10-ml shell
vial containing 3 ml of deionized water fortified with Nolan and
Carriker's (1946) salt formulation at a concentration of 1 ml/
liter.   After 90 min of exposure to miracidia, each snail was
removed from the vial, rinsed with deionized water, and placed
together with other snails of the same strain in finger bowls containing 200 ml of deionized water fortified as above.   These snails
were fed lettuce *ad libitum*.

At intervals of 8, 22, 48, and 72 hr after exposure to miracidia,
five snails of each strain were removed from the bowls, dissected
free of their shells, and fixed for 24 hr in 2% glutaraldehyde in
Sörensen's buffer, pH 6.5.   Subsequently, each snail was cut into
three parts, rinsed four times in tap water, and infiltrated with
Holt's (1959) hypertonic gum sucrose for 1 day to facilitate
sectioning in the frozen state.

All sections were cut at 10 μm in a Slee cryostat, mounted on
coverglasses, and air-dried.   Both Gomori's (1950) lead nitrate
method and Barka and Anderson's (1962) hexazonium pararosanilin
coupler method were employed for the demonstration of acid phosphatase on every third section.   Those sections to which Gomori's
method was applied were incubated in the substrate medium for 2 hr
at 37°C.   The sections subjected to Barka and Anderson's method
were incubated in the substrate-containing medium for 2 hr at 22°C.
Sections incubated in media devoid of the specific substrates and
media to which 0.01 M of sodium fluoride was added served as the
controls.

III.   RESULTS

A.   ACID PHOSPHATASE IN ISOLATED CELLS

Acid phosphatase activity was consistently detected in cytoplasmic granules of isolated granulocytes of *B. glabrata* (Fig. 1).
This enzyme was not detectable in isolated hyalinocytes.   Thus, I
am convinced that acid phosphatase is an effective marker enzyme
for granulocytes.

B.   ACID PHOSPHATASE IN CELLS PARTICIPATING IN ENCAPSULATION

Granulocytes participating in the encapsulation of *S. mansoni*
mother sporocysts in the highly resistant 10-R2 strain of *B.*
glabrata contained high levels of acid phosphatase activity.   This

Abbreviations used in Figs. 1-4: AP, acid phosphatase activity in snail granulocytes; C, capsule (granuloma) comprised of granulocytes; S, intact or degenerating mother sporocyst.

Fig. 1. Acid phosphatase activity in isolated granulocyte of *Biomphalaria glabrata*, strain LU$_{as}$. (Gomori's lead nitrate).

Fig. 2. Few acid phosphatase-rich granulocytes in vicinity of mother sporocyst in headfoot region of *Biomphalaria glabrata*, strain 10-R2, 8 hr post-exposure to miracidia. (Barka & Anderson's hexazonium pararosanilin).

Fig. 3. Capsule surrounding disintegrated mother sporocyst in headfoot region of *Biomphalaria glabrata*, strain 10-R2, 22 hr post-exposure to miracidia. Note that the capsule is comprised of tightly packed, acid phosphatase-rich granulocytes. (Barka & Anderson's hexazonium pararosanilin).

Fig. 4. Acid phosphatase in granulocytes of *Biomphalaria glabrata*, strain 10-R2, comprising the reduced capsule surrounding disintegrated mother sporocyst and in granulocytes intermingled with cellular debris. (Barka & Anderson's hexazonium pararosanilin).

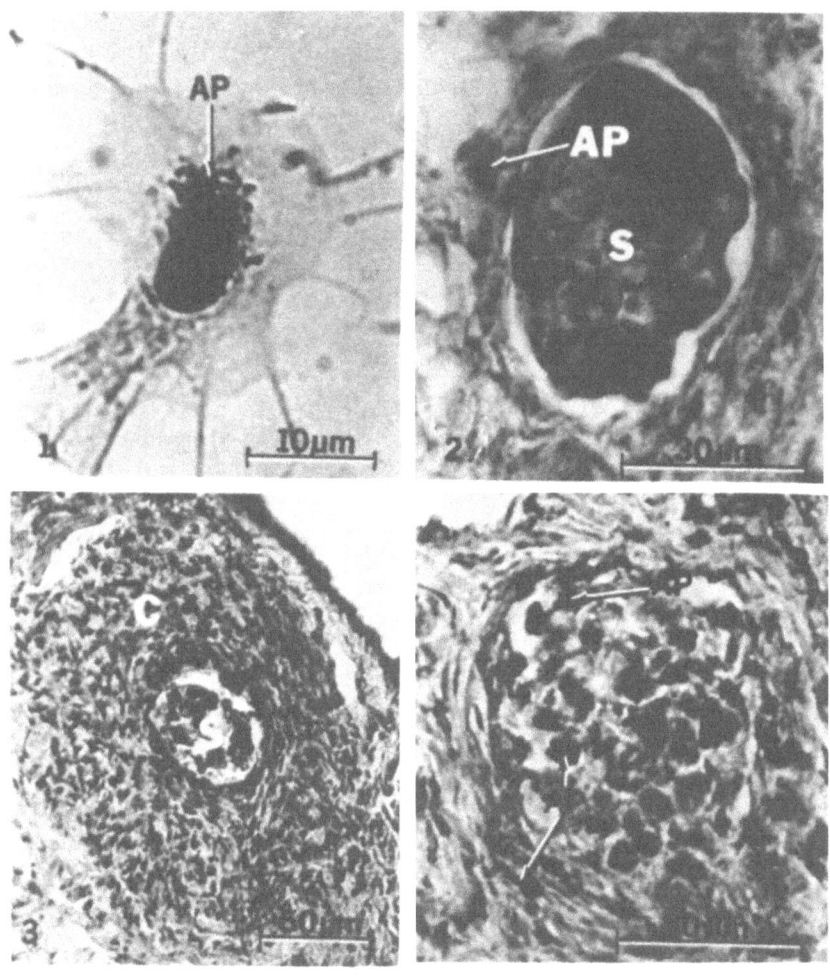

Abbreviations used in Figs. 5-8: AP, acid phosphatase activity
in snail granulocytes; S, mother sporocyst; SW, sporocyst wall.

Fig. 5. Acid phosphatase-rich granulocytes comprising the capsule
         surrounding disintegrated sporocyst in *Biomphalaria
         glabrata*, strain 10-R2, at 72 hr post-exposure to
         miracidia. (Barka & Anderson's hexazonium pararosanilin).

Fig. 6. Thin layer of flattened acid phosphatase-rich granulocytes
         surrounding viable sporocyst in headfoot of *Biomphalaria
         glabrata*, strain LU$_{bp}$, 48 hr post-exposure to miracidia.
         (Barka & Anderson's hexazonium pararosanilin).

Fig. 7. Acid phosphatase-rich granulocytes comprising capsule
         surrounding disintegrated sporocyst in headfoot of
         Biomphalaria glabrata, strain LU$_{bp}$, 48 hr post-exposure to
         miracidia. (Barka & Anderson's hexazonium pararosanilin).

Fig. 8. Acid phosphatase activity in wall of mother sporocyst of
         Schistosoma mansoni in headfoot of *Biomphalaria glabrata*,
         strain LU$_{as}$, 22 hr post-exposure to miracidia. Note
         absence of granulocytic capsule. (Barka & Anderson's
         hexazonium pararosanilin).

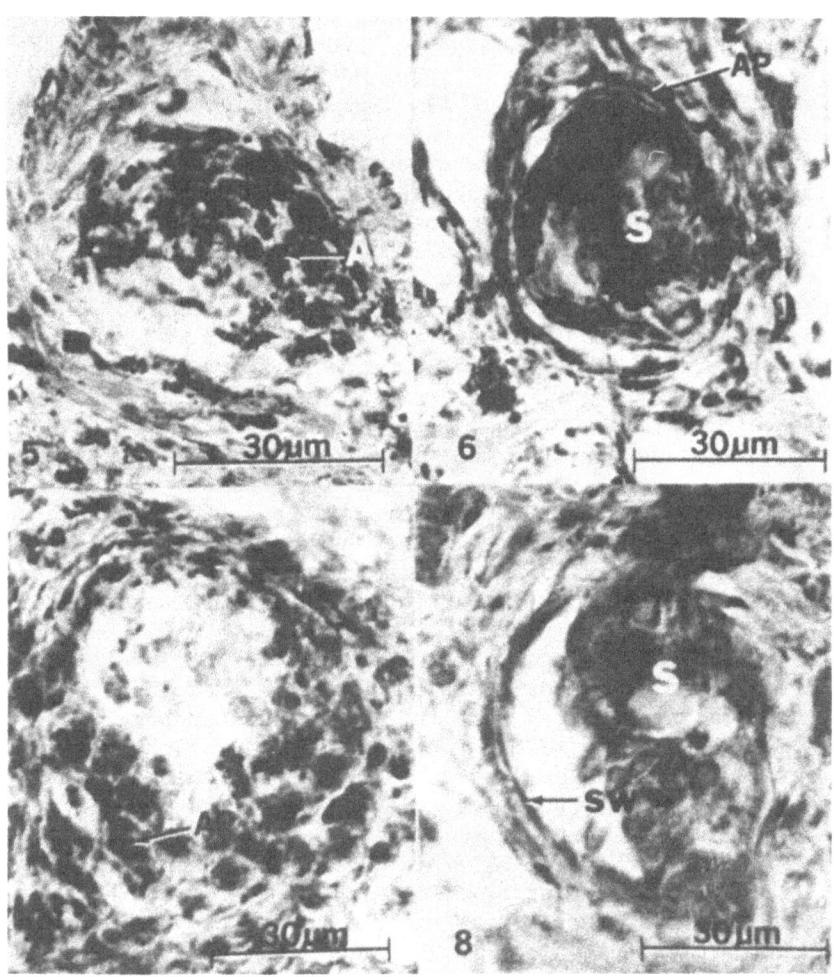

activity persisted throughout the encapsulation process. The
highest levels of acid phosphatase activity, as indicated by the
intensity of the reaction product, occurred in granulocytes which
remained at the sites of sporocyst encapsulation after the parasites or their remnants were no longer evident.

In strain 10-R2 snails, although there was some compression of
myofibers situated adjacent to young mother sporocysts in the
headfoot and tentacles, only a few isolated host granulocytes were
present on the surfaces of the parasites at 8 hr post-exposure to
miracidia (Fig. 2). These granulocytes contained acid phosphatase
activity, which clearly distinguished them from other types of
host tissues. At 22 hr post-exposure, tightly packed cellular
capsules, averaging 0.15 mm (0.075 - 0.275 mm) in diameter,
surrounded all mother sporocysts. Each granuloma was comprised
of only granulocytes, which were readily recognizable by the inclusion of acid phosphatase activity (Fig. 3). At 48 and 72 hr
post-exposure, the capsules were considerably smaller, averaging
0.08 mm (0.06 - 0.125 mm) in diameter. In most instances, little
or no recognizable schistosome material remained at the center of
each capsule in strain 10-R2 snails at 48 hr post-penetration
of miracidia. Rather, the loci at which the parasites were
originally situated were occupied by acid phosphatase-rich granulocytes and, in some instances, a few residual cell fragments of
the sporocysts (Fig. 4). By 72 hr post-penetration, each capsule
is comprised only of tightly packed acid phosphatase-rich granulocytes (Fig. 5).

Strain LU_{bp} snails exhibited variability in their responses to
our Puerto Rican *S. mansoni*. No host cellular responses to
mother sporocysts occurred at 8 and 22 hr post-exposure to
miracidia. However, by 48 hr post-exposure, a variety of host
cellular responses had developed. For example, in one snail all
mother sporocysts were not encapsulated and were developing
normally. However, three other snails contained granulocytic
capsules surrounding cellular debris. In the remaining snail of
this strain examined at 48 hr post-exposure, five foci of parasite
residency were observed. Three of the sporocysts were unencapsu-
lated and developing normally while the fourth was surrounded by
a thin layer of acid phosphatase-rich granulocytes but it appeared
viable (Fig. 6). The fifth sporocyst had disintegrated and the
residual cellular debris was surrounded by a thick granuloma, the
tightly packed granulocytes, of which it was comprised, were rich
in acid phosphatase (Fig. 7). At 72 hr post-exposure, there were
no granulocytes associated with surviving mother sporocysts.

In snails of strain LU_{as}, no encapsulation of mother sporocysts
occurred within 72 hr of exposure to miracidia. It is noted,
however, that in all instances, some acid phosphatase activity was
present in sporocyst walls (Fig. 8).

Isolated hemolymph cells and tissue sections incubated in media
without substrate or in media containing 0.01M sodium fluoride as
the inhibitor gave negative results for acid phosphatase.

IV. DISCUSSION AND CONCLUSIONS

It may be concluded from the results reported that granuloma formation as a reaction to *S. mansoni* sporocysts in incompatible strains of *B. glabrata* can be divided into two distinct stages. The first involves the accumulation of tightly packed acid phosphatase-rich granulocytes around the mother sporocyst. This stage is completed within 24 hr post-penetration by miracidia.

Death of the parasite marks the beginning of the second stage, which is characterized by the gradual disintegration and elimination of the sporocyst, and a decrease in the size of the granuloma. As the capsule decreases in size, acid phosphatase activity in the remaining granulocytes increases.

The pattern of granuloma formation described is consistently observed in totally resistant snails of the 10-R2 strain. A less consistent pattern of host response to larval *S. mansoni* of Puerto Rican origin occurs in *B. glabrata* of the partially resistant LU_{bp} strain. The pattern of host response in this strain is similar to that described by Newton (1954) in members of the F_2 generation resulting from a cross between susceptible and nonsusceptible parental strains. In some individuals of the LU_{bp} strain, encapsulation of all the mother sporocysts present occurred, while in others, encapsulation did not occur. As reported, in one snail, pronounced encapsulation, slight encapsulation, and nonencapsulation of sporocysts were observed. Newton (1954) has explained such variability in host response, particularly within an individual snail, as an expression of variability in the population of miracidia. On the other hand, Basch (1976), in his review, has concluded that compatibility between snail and schistosome is dependent upon each individual combination of snail and miracidium, although the factors involved remain essentially unknown.

Relative to the types of cells in *Biomphalaria* spp. involved in the encapsulation of helminth parasites, Brooks (1953) reported that both collagenous fibers and cells resembling fibroblasts comprised the capsule surrounding Puerto Rican *S. mansoni* miracidia in *B. obstructa*. Pan (1965) described the occasional encapsulation of degenerating *S. mansoni* mother sporocysts in susceptible *B. glabrata* to involve "fibroblasts" believed to have differentiated from hemocytes, and Carter and Bogitsh (1975) reported a few amoebocytes surrounding daughter sporocysts of *S. mansoni* in the digestive gland of a compatible strain of *B. glabrata*. In addition, Lie and Heyneman (1976) described the encapsulation of *Echinostoma lindoense* rediae in naturally resistant *B. glabrata* as an accumulation of hemolymph cells around the parasite. In an ultrastructural study on encapsulation of third-stage larvae of the nematode *Angiostrongylus cantonensis*, Harris (1975), based on morphological evidence, has concluded that the cells forming the granuloma are granulocytes.

By employing acid phosphatase as the marker, the evidence presented above indicates that the granulomas formed in nonsusceptible and partially susceptible strains of *B. glabrata* in response to mother sporocysts of *S. mansoni* are comprised solely of granulocytes. In this regard molluscan granulomas are simpler than mammalian granulomas; but then, there are fewer types of cells in *B. glabrata* (Cheng and Auld, 1977).

It is my opinion that true fibroblasts as found in vertebrates are not involved in the encapsulation of schistosome larvae in incompatible snails. This opinion is based on our finding that all of the cells comprising each capsule are rich in acid phosphatase. As Jeong and Heyneman (1976) and Yoshino and Cheng (1976) have demonstrated in isolated cells by employing cytochemistry at the fine structure level, this enzyme is restricted to the lysosomes of molluscan granulocytes. Also, Harris (1975) has reported that when examined with the electron microscope, what appeared to be "fibroblasts" comprising a capsule at the light microscope level were in fact elongate, pseudopod-producing granulocytes.

From the above, it is apparent that the study of granuloma formation in molluscs in response to such biotic invaders as larval schistosomes has much to offer towards understanding the mechanisms involved in this ubiquitous pathologic phenomenon. Specifically, (1) since the miracidia that were employed to infect LU_{bp} snails were genetically fairly homogenous, although not totally so, and yet there was such a diversity in the cellular response, it could be hypothesized, as had Newton (1954), Richards (1970,1973), and Richards and Merritt (1972), that granuloma development in resistant *B. glabrata* is genetically controlled. Hence, an understanding of the steps involved leading to the phenotypic expression from the genotype would appear to be of considerable interest.

(2) As stated earlier, according to Basch (1976), granuloma development is dependent upon each individual combination of snail and miracidium. This implies that the molluscan granulocytes recognize the incompatible schistosomes as foreign and the initial reaction cells, because of their apparently directional migration are probably chemotactically attracted to the parasites. This remains to be tested, and if found to be true, the helminth–mollusc model would appear to be a good one for further probing into the nature of the attractant. Such studies would be exceedingly interesting since in vertebrates it is known that chemotaxis is mediated by components of complement, primarily fragments of C_3 and C_5 and the macromolecular complex $C\overline{567}$ (Ward, 1975; and others). On the other hand, molluscs (and all other invertebrates) do not have complement. Therefore, it is hypothesized that another chemotactic mechanism must be operative.

(3) Also, it is not known whether cells that subsequently contribute to each granuloma is attracted to the initial cells, their products, the parasite, or some combination. Again, the molluscan model appears to be an ideal one for studying this aspect of granuloma formation.

(4) Finally, what is the role of the accumulated acid phosphatase in the granulomatous cells? Some speculations of this are presented in another contribution (Cheng, 1978).

In summary, it would appear that molluscs are useful models for studying nonimmunoglobulin and noncomplement mediated granuloma formation.

REFERENCES

Barka, T. and Anderson, P. J. (1962). Histochemical methods for acid phosphatase using hexazonium pararosanilin as coupler. *J. Histochem. Cytochem.*, 10, 741-753.

Basch, P. F. (1976). Intermediate host specificity in *Schistosoma mansoni*. *Exptl. Parasitol.*, 39, 150-169.

Brooks, C. P. (1953). A comparative study of *Schistosoma mansoni* in *Tropicorbis havanensis* and *Australorbis glabratus*. *J. Parasitol.*, 39, 159-165.

Carter, O. S. and Bogitsh, B. J. (1975). Histologic and cytochemical observations of the effects of *Schistosoma mansoni* on *Biomphalaria glabrata*. *Ann. N.Y. Acad. Sci.*, 266, 380-393.

Cheng, T. C. (1978). The role of lysosomal hydrolases in molluscan cellular response to immunologic challenge. *Comp. Pathobiol.*, This volume.

Cheng, T. C. and Auld, K. R. (1977). Hemocytes of the pulmonate gastropod *Biomphalaria glabrata*. *J. Invert. Pathol.*, 30, 119-122.

Cheng, T. C. and Yoshino, T. P. (1976). Lipase activity in the hemolymph of *Biomphalaria glabrata* (Mollusca) challenged with bacterial lipids. *J. Invert. Pathol.*, 28, 143-146.

DeDuve, C. (1969). The lysosome in retrospect. *In:* "Lysosomes in Biology and Pathology" Vol. 1. (J. T. Dingle and H. B. Fell, eds.). pp. 3-40. North Holland, Amsterdam.

DeDuve, C., Pressman, B. C., Gianetto, R., Wattiaux, R., and Appelmans, F. (1955). Tissue fractionation studies 6. Intracellular distribution patterns of enzymes in rat-liver tissue. *Biochem. J.*, 6, 604-617.

Domingo, E. O. and Warren, K. S. (1968a). The inhibition of granuloma formation around *Schistosoma mansoni* eggs. II. Thymectomy. *Am. J. Pathol.*, 51, 757-767.

Domingo, E. O. and Warren, K. S. (1968b). The inhibition of granuloma formation around *Schistosoma mansoni* eggs. III. Heterologous antilymphocyte serum. *Am. J. Pathol.*, 52, 613-631.

Domingo, E. O., Cowan, R. B. T., and Warren, K. S. (1967).
    The inhibition of granuloma formation around *Schistosoma
    mansoni* eggs. I. Immunosuppressive drugs. *Am. J. Trop.
    Med. Hyg.*, 16, 284-292.
Gomori, G. (1950). An improved histochemical technic for acid
    phosphatase. *Stain Technol.*, 25, 81-85.
Harris, K. R. (1975). The fine structure of encapsulation in
    Biomphalaria glabrata. *Ann. N.Y. Acad. Sci.*, 266, 446-464.
Harris, K. R. and Cheng, T. C. (1975). The encapsulation
    process in *Biomphalaria glabrata* experimentally infected
    with the metastrongylid *Angiostrongylus cantonensis*:
    light microscopy. *Intl. J. Parasitol.*, 5, 521-528.
Holt, S. J. (1959). Factors governing the validity of staining
    methods for enzymes, and their bearing upon Gomori acid
    phosphatase technique. *Exp. Cell Res.*, *Suppl.*, 7, 1-27.
Jeong, K. H. and Heyneman, D. (1976). Leukocytes of *Biomphalaria
    glabrata*: morphology and behavior of granulocytic cells
    in vitro. *J. Invert. Pathol.*, 28, 357-362.
Lie, K. J. and Heyneman, D. (1976). Studies on resistance in
    snails. 5. Tissue reactions to *Echinostoma lindoense* in
    naturally resistant *Biomphalaria glabrata*. *J. Parasitol.*,
    62, 292-297.
Lim, H. K. and Heyneman, D. (1972). Intramolluscan inter-
    trematode antagonism: a review of factors influencing the
    host parasite system and its possible role in biological
    control. *Adv. Parasitol.*, 10, 192-253.
Newton, W. L. (1952). The comparative tissue reaction of two
    strains of *Australorbis glabratus* to infection with *Schisto-
    soma mansoni*. *J. Parasitol.*, 38, 362-366.
Newton, W. L. (1954). Tissue response to *Schistosoma mansoni*
    in second generation snails from a cross between two strains
    of *Australorbis glabratus*. *J. Parasitol.*, 40, 352-355.
Newton, W. L. (1955). The establishment of a strain of
    Australorbis glabratus which combines albinism with high
    susceptibility to infection with *Schistosoma mansoni*. *J.
    Parasitol.*, 41, 526-528.
Nolan, L. E. and Carriker, M. R. (1946). Observations on the
    biology of the snail *Lymnaea stagnalis apressa* during twenty
    years in laboratory culture. *Am. Midl. Nat.*, 36, 467-493.
Novikoff, A. B. (1963). Lysosomes in the physiology and
    pathology of cells: contributions of staining methods.
    In: "Lysosomes" (A.V.S. de Reuck and M.P. Cameron, eds.).
    pp. 36-73. Little Brown, Boston, Mass.
Pan, C. T. (1965). Studies on the host-parasite relationship
    between *Schistosoma mansoni* and the snail *Australorbis
    glabratus*. *Am. J. Trop. Med. Hyg.*, 14, 931-976.
Perrotto, J. L. and Warren, K. S. (1969). The inhibition of
    granuloma formation around *Schistosoma mansoni* eggs. IV. X-
    irradiation. *Am. J. Pathol.*, 56, 279-291.

Raybourne, R. and Soloman, G. B. (1975). *Capillaria hepatica:* granuloma formation to eggs. III. Anti-immunoglobulin augmentation and reagin activity in mice. *Exp. Parasitol.*, 38, 87-95.

Raybourne, R. B., Solomon, G. B., and Soulsby, E. J. L. (1974). *Capillaria hepatica:* granuloma formation to eggs. II. Peripheral immunological responses. *Exp. Parasitol.*, 36, 244-252.

Richards, C. S. (1970). Genetics of a molluscan vector of schistosomiasis. *Nature*, 227, 806-810.

Richards, C. S. (1973). Susceptibility of adult *Biomphalaria glabrata* to *Schistosoma mansoni* infection. *Am. J. Trop. Med. Hyg.*, 22, 748-756.

Richards, C. S. (1975). Genetic factors in susceptibility of *Biomphalaria glabrata* for different strains of *Schistosoma mansoni*. *Parasitology*, 70, 231-241.

Richards, C. S. and Merritt, J. W. (1972). Genetic factors in the susceptibility of juvenile *Biomphalaria glabrata* to *Schistosoma mansoni* infection. *Am. J. Trop. Med. Hyg.*, 21, 425-434.

Solomon, G. B. and Soulsby, E. J. L. (1973). Granuloma formation to *Capillaria hepatica* eggs. I. Descriptive definition. *Exp. Parasitol.*, 33, 458-467.

Ward, P. A. (1975). Chemotactic factors of host origin. *Microbiology*, 1975, 200-201.

Warren, K. S. (1969). The inhibition of granuloma formation around *Schistosoma mansoni* eggs. V. "Hodgkin's-like lesion" in SJL/J mice. *Am. J. Pathol.*, 56, 293-303.

Warren, K. S. and Domingo, E. O. (1970). Granuloma formation around *Schistosoma mansoni, S. haematobium,* and *S. japonicum* eggs. Size and rate of development, cellular composition, cross-sensitivity, and rate of egg destruction. *Am. J. Trop. Med. Hyg.*, 19, 292-304.

Warren, K. S. and Kellermeyer, R. W. (1968). The foreign-body granuloma as a response to chemical mediators of inflammation. *J. Clin. Invert.*, 47, 99a.

Warren, K. S., Domingo, E. O., and Cowan, R. B. T. (1967). Granuloma formation around schistosome eggs as a manifestation of delayed hypersensitivity. *Am. J. Pathol.*, 51, 735-756.

Yoshino, T. P. and Cheng, T. C. (1976). Fine structural localization of acid phosphatase in granulocytes of the pelecypod *Mercenaria mercenaria*. *Trans. Am. Microsc. Soc.*, 95, 215-220.

INNATE CELLULAR DEFENSE BY MOSQUITO HEMOCYTES

DAVID A. FOLEY

Division of Parasitology
Department of Microbiology
New York University Medical School
New York City, New York

I. INTRODUCTION

The mosquito is an ideal laboratory model for the study of
innate cellular defense mechanisms among the Metozoa. Although
there may be nonspecific protective humoral factors present in
the hemolymph of mosquitoes, there is presumably no specific
humoral or antibody system as part of their arsenal of internal
defense. In addition, both man and mosquitoes are susceptible
to infection by parasites of the genus *Plasmodium*. In both
mammals and mosquitoes, the malaria sporozoite must migrate
through the circulation to reach a target organ (liver or salivary
gland), and is therefore potentially susceptible to the cellular
defense mechanisms of both hosts. This fortunate circumstance
affords an opportunity to study the response of a purely cellular
internal defense mechanism to the same stage of the same parasite
which successfully evades the defense mechanisms of both mammals
and mosquitoes.

Compared to many more popular laboratory animals, mosquitoes are
small and difficult to manipulate. However, the unique advantages
for studying innate cellular defense mechanisms presented by a
mosquito-*Plasmodium*-mammal system far outweigh the minor diffi-
culties of technique.

In this report, I would like to concentrate on the hemocytes of
mosquitoes, specifically on their role in internal defense, and
on their interactions with the developing malaria parasite. It
is not my intention to include an extensive review on the hemocytes
of insects or on cellular defense mechanisms in insects. These
topics have been the subject of several recent reviews (Arnold,
1974; Whitcomb *et al.*, 1974). Likewise, I do not intend to
discuss at length the phenomena of resistance and susceptibility
of mosquitoes to plasmodial infections, except as these relate
directly to cellular defense mechanisms in the mosquito.

Why are susceptible mosquitoes not resistant to plasmodial
infections, while they are apparently resistant to many other
potential microbial invaders? One would hope that the answers to
this very broad question might provide some insight into the work-
ings of innate cellular defense mechanisms in other model systems,
such as that of mammals, in which the reactivity of this mechanism
may be difficult to observe dissected away from the humoral response.

In my experimental approach to this problem, I have attempted
to determine (1) what types and how many hemocytes are present
in mosquitoes, (2) whether or not these hemocytes will interact
with foreign materials *in vivo* and *in vitro*, and (3) whether or
not these hemocytes will interact with sporogonic stages of the
malaria parasite.

A. *MOSQUITO REARING*

Adult female *Anopheles stephensi* mosquitoes were used in the
original studies reported here. Larvae were raised in an insectary

maintained at 25°C and 70% relative humidity. Adult mosquitoes were maintained at 18-21°C and 70% relative humidity. Developing larvae were fed finely ground laboratory chow (Ralston Purina Co.) supplemented with 10% liver powder (Nutritional Biochemical Corp.). Unless otherwise indicated, adult mosquitoes were maintained on 10% sucrose.

B. INFECTION OF MOSQUITOES WITH THE MALARIA PARASITE

Adult female mosquitoes 3 days post-emergence from pupae were infected with the NK-65 strain of *Plasmodium (berghei) berghei,* by permitting them to feed upon laboratory-reared golden hamsters with from 1-4% of their erythrocytes parasitized with gametocytes.

The malaria parasite was maintained by alternation of blood passages with the sexual cycle in the mosquito. The parasite was usually passed through the mosquito after three serial blood passages so that sufficient gametocytemias were maintained.

Hamsters on which mosquitoes were fed were injected intra-peritoneally (ip) with 0.1 ml of parasitized blood from a highly infected donor (parasitemia 40%) 3-4 days prior to mosquito feeding. Further details concerning this host parasite system were published by Vanderberg (1977).

II. DEVELOPMENT OF QUANTITATIVE METHODOLOGY

In order to study the hemocytes of *Anopheles stephensi,* it was first necessary to develop quantitative techniques for manipulation of these cells in the laboratory. These quantitative methods included: (1) development of a method for reproducible, accurate and rapid injection of materials into the mosquito hemocoel, (2) determination of mosquito hemolymph volume, and (3) development of a perfusion method for removing hemocytes from the hemocoel for observation and studies *in vitro*.

A. MICROINJECTION OF MOSQUITOES

A relatively atraumatic, rapid, and reproducible method of injection of mosquitoes was needed for several basic experiments including determination of hemolymph volume, quantification of hemocytes, and studies concerned with cellular defense *in vivo*.

Although a technique for quantitative microinjection of mosquitoes had been previously reported (Chao and Ball, 1956), this technique employed the use of a micromanipulator, which is tedious and time consuming to operate. The method also required special spindle-shaped glass capillary needles for quantitative injection. Because of these complications, I decided not to use the Chao and Ball (1956) method of microinjection, but to develop a rapid, more simple method.

This was accomplished in the following way. Fine microinjection
needles with a relatively constant bore were drawn from glass
capillary tubes. These glass capillaries were then fitted to a
capillary holder (Leitz) and a rubber tube with a mouthpiece so
that positive pressure could be applied to force liquids through
the needle. The capillary holder, and thus the needle, were
manipulated by hand. The amount of liquid expelled from the
needles was determined by measuring the distance travelled by
the liquid meniscus down the capillary during the injection
procedure. The needles were calibrated using radioisotope so
that the volume injected could be calculated from the movement
of the meniscus. Needles were calibrated by expelling a volume
of ^{14}C-inulin solution equivalent to 1 mm of motion of the meniscus in the capillary onto discs of filter paper. The discs
were placed in scintillation vials with scintillation cocktail
and counted on a Searle Isocap scintillation counter. The average
volume injected for 10 such needles was 0.204µl, with a standard
deviation of 0.015µl (7.6%).

Injection of mosquitoes was visually monitored by the use of a
dissection microscope, but no micromanipulator was needed to hold
the mosquito or the needle. Mosquitoes to be injected were first
chilled to anaesthetize them, and were then placed on a glass
plate which was in contact with a block of ice. A millimeter
ruler was also placed on the glass plate in such a position as
to be easily visible in the field of the dissection. Mosquitoes
were manipulated into position with the fine forceps. The
injection needle was placed into the right or left thorax, and
the needle was then placed parallel to the ruler so that the
distance of movement of the meniscus in the needle could be
observed accurately. Following injection, mosquitoes were gently
removed from the needle and brought to room temperature.

This method of injection was suitable by several criteria. It
was rapid. Over 100 mosquitoes could be injected in an hour
by a single operator. It was reproducible and accurate (see
above). It produced very little mortality. When Tissue Culture
Medium (M199, Gibco) was injected alone into 25 mosquitoes, the
mortality was only 20% at the end of 2 weeks.

To determine the immediate distribution of injected fluids,
a dye (0.01% Trypan blue in M199) was injected into mosquitoes.
Although 1% Trypan blue killed 23 out of 26 mosquitoes by the
second day post-injection, 0.01% Trypan blue was well tolerated,
and after 8 days post-injection, none of 27 mosquitoes had died.
In mosquitoes being injected with 0.01% Trypan blue, the dye
could be seen spreading throughout the abdomen and thorax. Therefore, it is reasonable to assume that injected fluids are
initially well distributed throughout at least the two main body
cavities of the mosquito.

In other experiments, it would be necessary to inject mosquitoes with suspensions of sporogonic stages of the malaria parasite that contained microbial contaminants, so the maximum tolerated dosages of several antibiotics were determined for possible prophylactic use. Gentamycin (Schering Corporation), Kanamycin (Bristol Laboratories), and a mixture of Penicillin, Streptomycin, and Fungizone (GIBCO) were diluted in M199 and groups of 10-15 mosquitoes were injected with 0.20µl of each dilution of each antibiotic. Mosquito mortality was determined each day for 2 weeks. The maximum tolerated dosage was determined as that dosage which produced the lowest mortality over the two weeks post-injection. The results of this experiment are given in Table 1.

B. MOSQUITO HEMOLYMPH VOLUME

In order to ascertain the dimensions of the theater in which internal defense based on hemocytes occurs, and to be able to calculate the density of these hemocytes in their fluid environment, it was necessary to know the hemolymph volume of the mosquito.

The determination of the hemolymph volume was a difficult problem due to the small size of the mosquitoes, and the negative hemolymph pressure in the adult mosquitoes (Jones, 1954). To accomplish this aim, a centrifugation method used for collecting hemolymph from other insects for protein studies was adapted to a radiodilution technique for hemolymph volume determination (Mack *et al.*, 1978). Because of the small volumes of hemolymph obtainable from individual mosquitoes by this method, 50 mosquitoes were processed together for each experiment. In the centrifugation procedure used to remove hemolymph from the mosquitoes, groups of 50 mosquitoes were anaesthetized by cooling to 4°C. The tips of the prosci of anaesthetized mosquitoes were cut off, and a small hole was cut in the distal abdomen of each mosquito. The 50 mosquitoes were then placed in a small conical centrifuge tube over a plug of glass wool, and centrifuged at 750 g for 10 min. The hemolymph that collected in the bottom of the tube was measured with a 10µl Hamilton syringe. Volumes of hemolymph collected from mosquitoes of different ages and in the malaria infected and noninfected condition are presented in Table 2. From these data it is apparent that increase in age post-emergence is associated with a decrease in the amount of hemolymph that may be obtained by this method. This hemolymph collection procedure was used as part of the radioisotope dilution method described below for the determination of hemolymph volume.

Table 1. Maximum Tolerated Dosages of Antibiotic Solutions
         Injected into *Anopheles stephensi* Mosquitoes.
         Injection Volume = 0.20 Microliter/Mosquito

Antibiotic Solution	Maximum Dosage
A Penicillin	100 U/ml
Streptomycin	100 mcg/ml
Fungizone	0.25 mcg/ml
B Gentamycin	0.50 mg/ml
C Kanamycin	10 mg/ml

Table 2. Volume of Hemolymph Collected From *Anopheles stephensi* Mosquitoes by the Centrifugation Method.[a]

Age of Mosquitoes (Days Post-Emergence)	Type of Blood Meal[b]	Number of Experiments[c]	Volume of Hemolymph Recovered. (nl/Mosquito + Standard Deviation)
0	None	23	110+15
7	Non-infected	13	90+10
	Infected	12	80+4
14	Non-infected	25	80+9
	Infected	22	80+9
21	Non-infected	13	70+7
	Infected	11	70+5

a   After Mack *et al.* (1978).

b   At 3 days post-emergence, mosquitoes were permitted to feed upon non-infected or
Plasmodium berghei infected hamsters.

c   Fifty female mosquitoes were used in each experiment.

A buffered salt solution (145mM NaCl, 20 mM KCl, 2.5 mM
$CaCl_2 \cdot 2H_2O$, 1.0 mM $MgCl_2 \cdot 6H_2O$ and 25 mM PIPES buffer sodium salt,
pH 6.9, osmolality 400 mOsm; Mack *et al.*, 1978), containing ^{14}C-
carboxyinulin (specific activity 1.82 mCi/g) was injected into
cold, anaesthetized mosquitoes by the microinjection method
described above. Each mosquito received 0.19-0.22µl of the
diluted isotope. The mosquitoes were then permitted to regain
activity at room temperature (21-23°C) for 20 to 30 min so
that adequate mixing of hemolymph and isotope could occur.
Hemolymph was removed from these mosquitoes by the centrifugation
method described above. The recovered hemolymph was measured
with a 10µl Hamilton syringe and transferred to scintillation
vials for counting. In order to ascertain the amount of isotope
remaining in the mosquitoes, the bodies of the mosquitoes were
homogenized and placed into scintillation vials for counting.
The amount of isotope remaining associated with the glass wool
and centrifuge tube was determined by rinsing these with 2 ml
of distilled water, which was subsequently added to scintillation
vials. The total radioactivity of each sample was measured in
a Searle Isocap 300 liquid scintillation counter with external
quench correction. All samples were counted at the 0.25% level
of confidence (200,000 counts) and recounted to exclude chemi-
luminescence. Hemolymph volume was calculated from the amount
of isotope dilution which had occurred in the hemolymph.

The hemolymph volumes of mosquitoes are presented in Table 3.
These volumes represent the average hemolymph volume in individual
mosquitoes in experimental groups of 50, at different ages, and
in the infected and noninfected states. The largest hemolymph
volumes were found in newly emerged mosquitoes. Increasing age
post-emergence was associated with a decrease in total hemolymph
volume. However, malaria infected and noninfected mosquitoes
had approximately the same hemolymph volumes.

A comparison of the hemolymph volume determined by the radioisotope dilution technique (Table 3) with the volume of hemolymph
obtainable from mosquitoes by the centrifugation method (Table 2)
demonstrates that there is a considerable amount of hemolymph
left in mosquitoes following centrifugation. However, the
decrease in volume measured in both ways is associated with
increasing age of adult mosquitoes and suggests that there is
a real decrease in the hemolymph volume of mosquitoes as they
increase in age. This may be due to the relative increase in
these female mosquitoes of the fat body and ovaries.

C. QUANTITATIVE REMOVAL OF HEMOCYTES FROM THE MOSQUITO HEMOCOEL

In order to perform quantitative cytological studies with
mosquito hemocytes, it was necessary to develop a technique for
quantitatively collecting hemocytes from the hemocoels of individual mosquitoes. This was accomplished by the adaptation of a

Table 3. Hemolymph Volume of Female *Anopheles stephensi* Mosquitoes
          Determined by the Radioisotope Dilution Technique[a].

Age of Mosquitoes (Days Post-Emergence)	Type of Blood Meal[b]	Hemolymph Volume (nl/Mosquito)
0	None	347
0	None	383
0	None	304
0	None	309
3	None	310
14	Non-infected	190
14	Infected	183

a After Mack, Foley, and Vanderberg (1978).

b At 3 days post-emergence, mosquitoes were permitted to feed
   upon non-infected or *Plasmodium berghei* infected hamsters.

method developed for studying hemolymph proteins in *Aedes*
aegypti by Dr. Andy Spielman (pers. comm.). With this procedure,
the hemocoel of each mosquito was perfused with 100μl of buffered
salt solution (BSS) or M199, into tubes of fixative or onto
slides for direct observation. By this means, hemocytes which
were free in the hemolymph would then be perfused out of the
hemocoel. Fine needles were drawn by hand from the tips of
disposable Pasteur pipettes. These needles were filled with
100μl of M199 or BSS, and the fine needle point inserted into
the thorax of a chilled, anaesthetized mosquito. The mosquito
was immobilized on the pipette tip with a drop of cement and a
small hole was cut in the distal abdomen with iridectomy scissors.
Positive gas pressure was used to perfuse the solution through
the mosquito and out of the hole in the abdomen.

This perfusion method was quantitated so that the amount of
the hemocoel that was perfused could be calculated. Mosquitoes
were injected with ^{14}C-carboxyinulin by the method described
above, and the amount which was perfused out by the perfusion
method was determined. The data from these calibration experiments are presented in Table 4. The average portion of the
hemocoel perfused during these experiments was 59.2%, with a
standard deviation of 14.8%. Therefore, using this calibrated
perfusion method, as well as counts of the hemocytes present in
the perfusate, it was possible to calculate the total number of
hemocytes in the mosquito hemocoel. A major presumption of this
method is that the hemocytes in all parts of the hemocoel are
equally accessible to removal by perfusion and that these hemocytes are evenly distributed throughout the hemocoel.

III. HEMATOLOGY OF *ANOPHELES STEPHENSI*

A. *LIVING HEMOCYTES*

Hemocytes were removed from female mosquitoes by the perfusion
method for direct observation or for counting. For observations
of living hemocytes, hemolymph was perfused directly onto clean
glass slides and covered with a cover glass, which was sealed
in place with petroleum jelly. Hemocytes in these preparations
were observed with phase-contrast microscopy.

Based upon morphological criteria, I have tentatively placed
the hemocytes of *Anopheles stephensi* into five classes according
to the well-established system of classification of insect blood
cells developed by Jones (1962,1970). The hemocyte types
observed by me from *A. stephensi* include prohemocytes, plasmatocytes, granular hemocytes, adipohemocytes, and spherule cells.

Observations on the behavior of living hemocytes revealed that
the prohemocytes, plasmatocytes, and granular hemocytes were
motile, and were capable of adhering to and spreading upon the

Table 4. Determination of the Efficiency of Hemocoel Perfusion of *Anopheles stephensi* Mosquitoes by a Radioisotope Method.[a]

Experiment	Age of Mosquitoes (Days Post-Emergence)	Blood Meal[b]	Percentage of Hemocoel Perfused[c]
1	0	None	47
2	0	None	52
3	0	None	51
4	0	None	50
5	0	None	52
6	1	None	85
7	2	None	80
8	11	Infected	47
9	11	Non-infected	69

a Mosquitoes were injected with ^{14}C inulin and subsequently perfused with 100µl of M199 or BSS.

b At 3 days post-emergence, mosquitoes were permitted to feed upon non-infected or *Plasmodium berghei* infected hamsters.

c Expressed as the percentage of injected isotope that was perfused out of the mosquitoes. n=8 mosquitoes/experiment.

Figs. 1-4.   Photomicrographs of living hemocytes of *Anopheles*
             stephensi observed with phase-contrast optics.
             N = nucleus.

             Fig. 1.   Prohemocyte
             Fig. 2.   Plasmatocyte
             Fig. 3.   Granular hemocyte
             Fig. 4.   Adipohemocyte

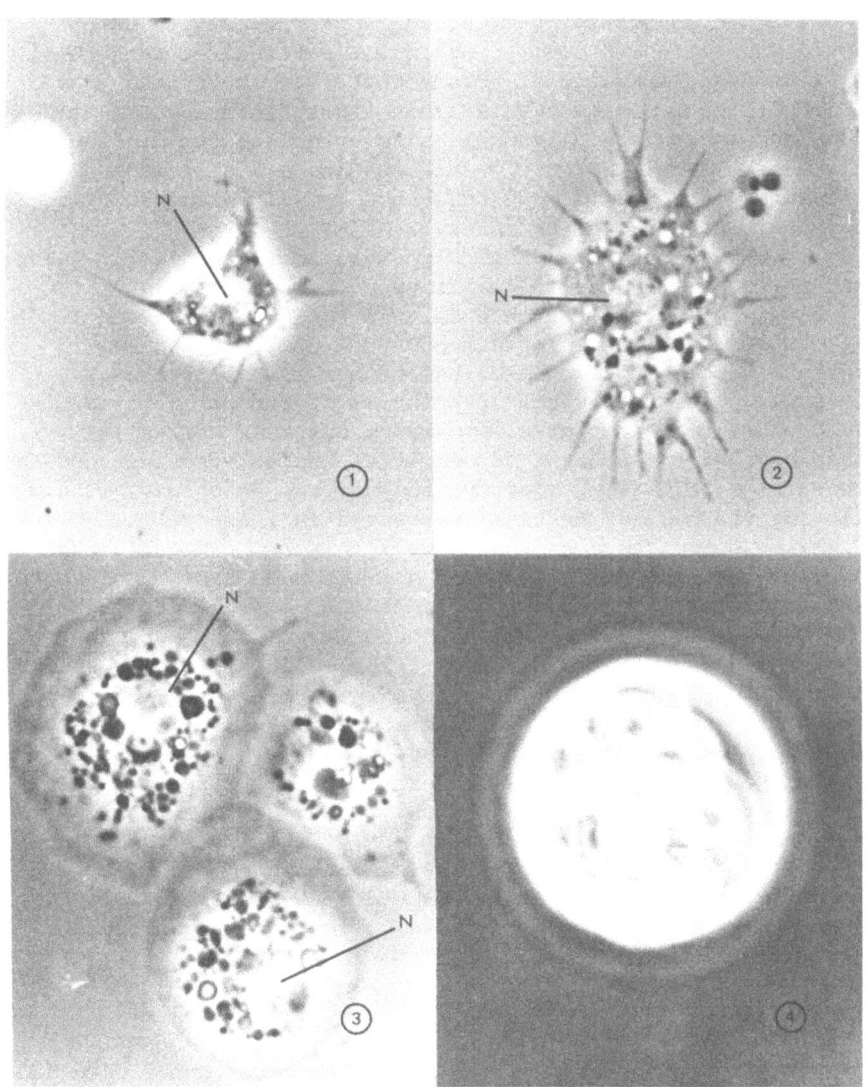

glass substrate. Spreading of these cells was initiated as soon
as they were perfused onto the glass substrate. Spherule cells
and adipohemocytes were not motile and did not attach to glass.
The morphology of prohemocytes, plasmatocytes, and granular
hemocytes was best observed when these cells were spread over
the substrate. Freshly obtained cells of these types were
rounded and quite refractile, making it difficult to observe
the cytoplasmic inclusions. Prohemocytes in the spread state
in vitro (Fig. 1) possess relatively few striking cytoplasmic
inclusions and have a high nucleus to cytoplasm ratio. The
nucleus is slightly eccentric and contains a single large
nucleolus. There may be one to several small filopodia or
pseudopodia at the cell periphery.

Plasmatocytes spread extensively *in vitro* (Fig. 2). The endoplasm of each cell contains small vacuoles and granules of
various sizes, as well as an eccentric nucleus with a single
large nucleolus. The ectoplasm of these cells is not well
demarcated from the endoplasm, but rather gradually changes
to a nongranular peripheral cytoplasm. At the periphery of the
spread plasmatocyte, there are usually numerous spike-like
pseudopodia with a webbing of nearly agranular ectoplasm spread
between them (Fig. 2). Slow cyclosis of the endoplasmic inclusions
including the nucleus is easily observed in these cells.

Granular hemocytes also spread well *in vitro* (Fig. 3). The
endoplasm of these cells contains a large number of phase-dense
granules of different shapes and sizes, some small vacuoles,
and an eccentric nucleus with a single large nucleolus. The
ectoplasm is essentially agranular, formed into broad lamellipodia,
and may also possess some spike-like filopodia.

There are apparently morphological gradations between prohemocytes and plasmatocytes, as well as between plasmatocytes and
granulocytes. These involve increases in cell size when spread,
and relative numbers of granules and vacuoles (Figs. 1-3).

Adipohemocytes do not spread *in vitro* (Fig. 4). These rounded
cells which vary considerably in size contain numerous large
refractile droplets. Spherule cells disintegrate easily *in vitro*
and change rapidly from large round cells to disorganized and
amorphous cell fragments and extrusions.

B. DIFFERENTIAL AND TOTAL CELL COUNTS

Differential counts as well as total hemocyte counts were made
of hemocytes from *Plasmodium*-infected as well as noninfected
mosquitoes of different ages (Table 5). Total cell counts were
performed by perfusing hemocytes directly into 2.5% glutaraldehyde in BSS, and counting them in a hemacytometer. These counts
were corrected for dilution by the perfusion solution and the
fixative, as well as for the percentage of the hemocoel which

Table 5. Hematological Parameters of *Anopheles stephensi*.

Age of Mosquitoes (Days Post-Emergence)	Blood Meal[a]	Number of Mosquitoes	Total Cell Count[b]	Differential Cell Counts[c]				
				PRO	PLA	GRA	ADI	SPH
0	None	10	X=10,220	5.8	64.4	15.5	9.0	5.3
			SD=3,848	3.4	12.0	6.6	3.3	3.1
1	None	10	X=	12.9	66.6	8.9	10.3	2.2
			SD=	6.4	12.0	6.0	6.9	3.3
2	None	10	X=9,262	--	--	--	--	--
			SD=4,291					
10	None	5	X=1,725	6.4	88.0	2.6	1.8	1.4
			SD=1,014	1.3	6.4	3.2	3.5	1.5
14	Infected	5	X=3,449	13.6	74.6	6.0	2.4	3.4
			SD=1,953	1.9	5.2	1.6	1.9	3.9

a   At 3 days post-emergence, mosquitoes were permitted to feed upon a hamster infected with *Plasmodium berghei*.

b   Hemocytes were obtained by perfusing the hemocoels of mosquitoes. The hemocyte counts shown here have been corrected for the percentage of the hemocoel perfused by this method.

c   PRO=Prohemocytes, PLA=Plasmatocytes, GRA=Granular hemocytes, ADI=Adipohemocytes, SPH=Spherule cells.

Figs. 5-8. Photomicrographs of hemocytes of *Anopheles stephensi*
            (glutaraldehyde fixed, Giemsa's stain).
            N = Nucleus.

            Fig. 5.   Prohemocyte
            Fig. 6.   Plasmatocyte
            Fig. 7.   Granular hemocyte
            Fig. 8.   Group of hemocytes

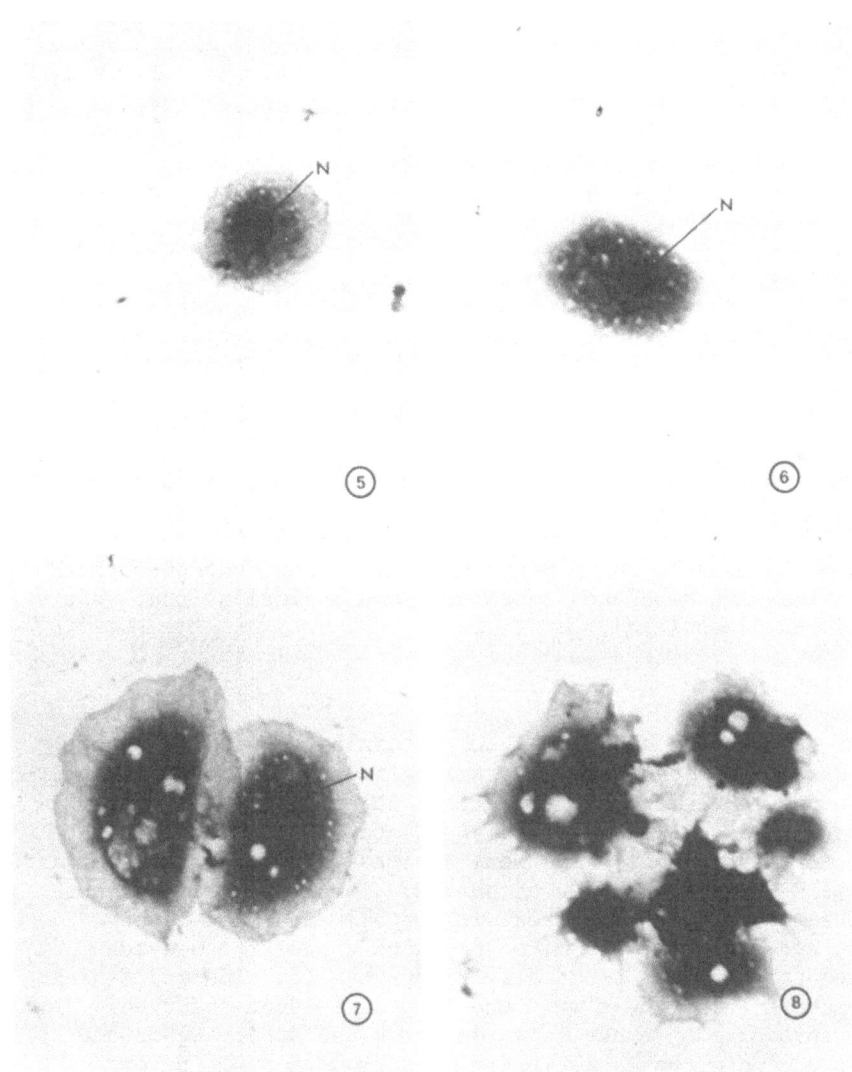

had been perfused, and thus represent the total number of hemocytes which could be perfused from an individual mosquito if it
were possible to perfuse 100% of the hemocoel. The largest
number of hemocytes were perfused from newly emerged mosquitoes,
while considerably less hemocytes were perfused from 10-day-old
mosquitoes (Table 5). It must be noted, however, that because
the 10-day-old mosquitoes had not received a blood meal, it is
possible that they were not nutritionally balanced. Over a long
period of time, such as 10 days, this nutritional imbalance
may have been expressed in the hematology of the mosquito. The
hemocyte count for mosquitoes which had received an infected
blood meal was roughly one-third that of newly emerged mosquitoes
but was about the same as that for the 10-day-old mosquitoes
that had received no blood meal. The differential counts of
hemocytes from malaria-infected mosquitoes were essentially the
same as that for noninfected ones. The most numerous hemocyte
class in all groups of mosquitoes was the plasmatocyte (Table 5).

C. VITAL STAINING

Hemocytes were observed in 0.002% solutions of Janus Green B
or neutral red in M199. Virtually all hemocytes possessed
inclusions that were intravitally stained by Janus Green B.
Adipohemocytes were not stained by neutral red, but the other
four classes of hemocytes contained some inclusions that were
stained by neutral red.

D. FIXED AND STAINED HEMOCYTES

Hemocytes were fixed with 2.5% glutaraldehyde in BSS after
various time periods *in vitro* ranging from 5 min to 30 min post-
perfusion. After 5 min of fixation, the slides were gently
rinsed with distilled water, dipped into 95% ethanol, and subsequently air dried. Cells were stained with 5% Giemsa's stain
in 0.01 M phosphate buffer at pH 7.0. These preparations were
examined mounted or unmounted with an oil immersion lens.
Only prohemocytes, plasmatocytes, and granular hemocytes
adhered to the glass substrate through the fixation and staining
procedure. Thus, these were the only cell types observed.
The spreading of these cells on the glass substrate became
more pronounced the longer the time period *in vitro* before
fixation. Those hemocytes that were fixed at 30 min post-perfusion
were spread extremely thinly and contained many vacuoles. A
spreading time of 10 min was selected to provide the best morphological and staining characteristics.
Prohemocytes (Fig. 5) contained a densely basophilic foamy
cytoplasm with very few particulate inclusions. There were few
to several small pseudopodia in which ectoplasm could not be
distinguished from endoplasm. The eccentric nucleus included
a single large acidophilic nucleolus.

The cytoplasm of plasmatocytes (Fig. 6) stained basophilically, but it was not as densely stained as that of most prohemocytes. A variable number of vacuoles, as well as acidophilic and basophilic inclusions were present, mostly in the central region of the spread cells. Around some plasmatocytes, there was an agranular ectoplasm which consisted of spike-like filopodia projecting radially past the cell periphery, with webs of ectoplasm between them. There were many more filopodia on these cells than on the prohemocytes. In some cells, the filopodia contained a blue faintly staining rod-like structure, which formed the axis of the filopodial projection. The nuclei of plasmatocytes stained very faintly pink with single, large, intensely acidophilic nucleoli.

Granular hemocytes (Fig. 7) were distinguishable by the large numbers of basophilic and acidophilic granules as well as vacuoles which were present within their endoplasms. The nuclei of these cells stained pink and contained a single acidophilic nucleolus. The ectoplasm of these cells was thinly spread and consisted of lamellipodia and/or filopodia.

Although most plastocytes and granular hemocytes were mononucleate, some binucleate or multinucleate representatives were occasionally observed, especially from newly-emerged mosquitoes. These multinucleate cells were morphologically similar to mononucleate hemocytes except that they were larger when spread, and contained more than one nucleus.

Hemocytes were arranged on the slide in these fixed and stained preparations in three types of conformations. Single hemocytes were the most prominant (Fig. 6), pairs of hemocytes together were the next most prominant (Fig. 7), while groups of three or more (frequently 8-12) were the least common (Fig. 8). Of 33 such conformations picked at random from one sample, 20 were single cells, 8 were pairs, and 5 were groups of three or more hemocytes.

E. HEMATOLOGY OF OTHER MOSQUITOES

There have been few other observations on the hematology of mosquitoes. Perhaps the most complete previous study is that of Amouriq (1960) on the changes in the hemocyte formula during the development of *Culex hortensis*. Amouriq found that the hemolymph of this mosquito always included proleucocytes, developing macronucleocytes, macronucleocytes, macronucleocytes with vacuoles, granulocytes, and granulocytes in senescence. The most numerous of these in larvae, pupae, and adult mosquitoes were the granulocytes in senescence. In addition to these, the pupae possessed fusiform hemocytes, and adults possessed micronucleocytes. Although Amouriq (1960) did not follow the classification system of Jones (1962), it seems likely from his descriptions that at

least the following are synonomous: (1) proleucocytes-prohemo-
cytes; (2) macronucleocytes=plasmatocytes; and (3) granulocytes=
granular hemocytes. Thus, adult *C. hortensis* and *A. stephensi*
have in common at least these three hemocyte types.

Larval *Aedes* mosquitoes also contain one or more of the hemocyte types found in adult *Anopheles* and *Culex* mosquitoes. In
larval *Aedes albopictus*, Bhat and Singh (1975) detected five
varieties of hemolymph cells: prohemocytes, plasmatocytes, two
types of podocytes, and spherules. In addition, in larval *Aedes
aegypti*, crystal cells (Bronskill, 1962) and spindle-shaped
plasmatocytes and oenocytoids (Andreadis and Hall, 1976) have
been reported. Jones (1954) observed two types of hemocytes in
larval *Anopheles quadrimaculatus*. These were most numerous in
late 4th instar larvae and in pupae and consisted of round to
ovoid cells with either granular or nongranular cytoplasm.

IV. FUNCTIONAL STUDIES

A. *HEMOCYTE INTERACTION WITH FOREIGN PARTICULATE MATERIALS*

1. *Studies in vitro*

The ability of the hemocytes of *A. stephensi* to interact
or associate *in vitro* with foreign particulate materials was
tested in the following way. The terms interaction and association are here used to describe the ability of the hemocyte
either to adhere directly to, or to phagocytose a foreign particle,
because it is difficult to determine whether or not a particle
is actually completely within a hemocyte. Hemocytes were perfused from the hemocoels of mosquitoes directly onto cleaned glass
slides, or into the individual chambers of Chamber Slides (Lab-
Tek). The hemolymph was permitted to stand in a moist chamber
at room temperature (21-23°C) for 5 min so that the hemocytes
would settle and attach to the glass. At this time, a suspension
of glutaraldehyde fixed human RBC's (GRBC), *Escherichia coli*, or
Bacillus subtilis, was added to the hemolymph on the slide, and
the preparations were permitted to stand an additional 15 min
at room temperature. The slides were then rinsed gently and
fixed with 2.5% glutaraldehyde in BSS, rinsed with distilled
water, rinsed with 95% ethanol, and air dried. Fixed preparations
were then stained with Giemsa's stain as described above.

Because only the prohemocytes, plasmatocytes, and granular
hemocytes adhere to the substrate under the conditions described,
these are the only cell types which it was possible to examine
by this method. However, for several reasons, it was not possible to differentiate between these cell types in these special
preparations. Briefly, it was necessary to leave the hemocytes
in vitro for a total of 20 min to permit interaction or association to occur, and during this lengthy time period, extensive

spreading also occurred. In addition, the cells which were
associated with foreign particles were intensely vacuolated, and
possibly degranulated as described for molluscan hemocytes in a
similar system by Foley and Cheng (1977). Furthermore, many of
the cells reacted so well with the test particles, that their
morphological characteristics were obscured by adhering test
particles. Those few cells that did not interact with test
particles were of all three types.

Hemocytes from noninfected mosquitoes interacted *in vitro* with
the three types of test particles. Over 95% of the hemocytes in
each type of preparation had attached to or phagocytosed one or
more GRBC (Fig. 9), *E. coli* (Fig. 10), or *B. subtilis* (Fig. 11).

Because it is possible that the *Plasmodium* infection in the
mosquito might cause a general suppression in the ability of the
mosquito hemocytes to respond to any foreign invader, the ability
of the individual hemocytes to respond *in vitro* to GRBC was
determined. Hemocytes from mosquitoes which had fed on infected
hamsters 11 days previously were perfused onto glass slides and
tested as described above for their ability to interact *in vitro*
with GRBC. Over 95% of these hemocytes associated with at least
one GRBC. Thus, hemocytes from these mosquitoes were not
deficient in their ability to recognize and interact *in vitro*
with GRBC when compared with hemocytes from noninfected mosquitoes.

2. *Studies in vivo*

Using the microinjection technique described above, suspensions of GRBC were injected into newly emerged mosquitoes. At
0,3, and 24 hr post-injection, mosquitoes were fixed, embedded in
Ester wax (BDH Chemicals, Ltd.), and sectioned at 7 to 10μm.
Sections were stained with hematoxylin and eosin or Giemsa's
stain.

In the mosquitoes that were fixed at 0 hr post-injection, the
GRBC were dispersed throughout the hemocoel of the thorax and
abdomen. In the thorax, there were some small aggregates of
GRBC which were in association with mosquito cells in pockets
around the thoracic musculature. In the mosquitoes fixed at 3
and 24 hr post-injection, most of the GRBC were sequestered into
aggregates of several GRBC, in various locations throughout the
hemocoel. In the 0 hr samples, the GRBC had been dispersed as
nearly all single cells. Therefore, a segregation of the GRBC
had taken place within 3 hr post-injection of the GRBC, and was
slightly more pronounced by 24 hr post-injection. There was a
moderate amount of association of the hemocytes in the vicinity
of the thoracic musculature, with clumps of GRBC (Fig. 12).
These cells possessed some morphological characteristics of
plasmatocytes or granular hemocytes. However, it must be noted
that many of the sequestered clumps of GRBC did not have a visible
association with mosquito hemocytes.

Figs. 9-12.   Photomicrographs of hemolymph cells of *Anopheles*
              stephensi which have interacted *in vitro* or *in vivo*
              with suspensions of foreign particles. (Figs. 9-11,
              glutaraldehyde fixed, Giemsa's stain; Fig. 12,
              glutaraldehyde fixed, hematoxylin and eosin).
              G = glutaraldehyde-fixed red blood cell, E =
              Escherichia coli, B = *Bacillus subtilis*, N =
              Hemocyte nucleus.

              Fig. 9.   Group of hemocytes in association with
                        glutaraldehyde-fixed red blood cells.
              Fig. 10. Hemocyte in association with heat-killed
                        Escherichia coli.
              Fig. 11. Hemocyte in association with heat-killed
                        Bacillus subtilis.
              Fig. 12. Hemocyte in mosquito hemocoel in association
                        with injected glutaraldehyde-fixed red blood
                        cells.

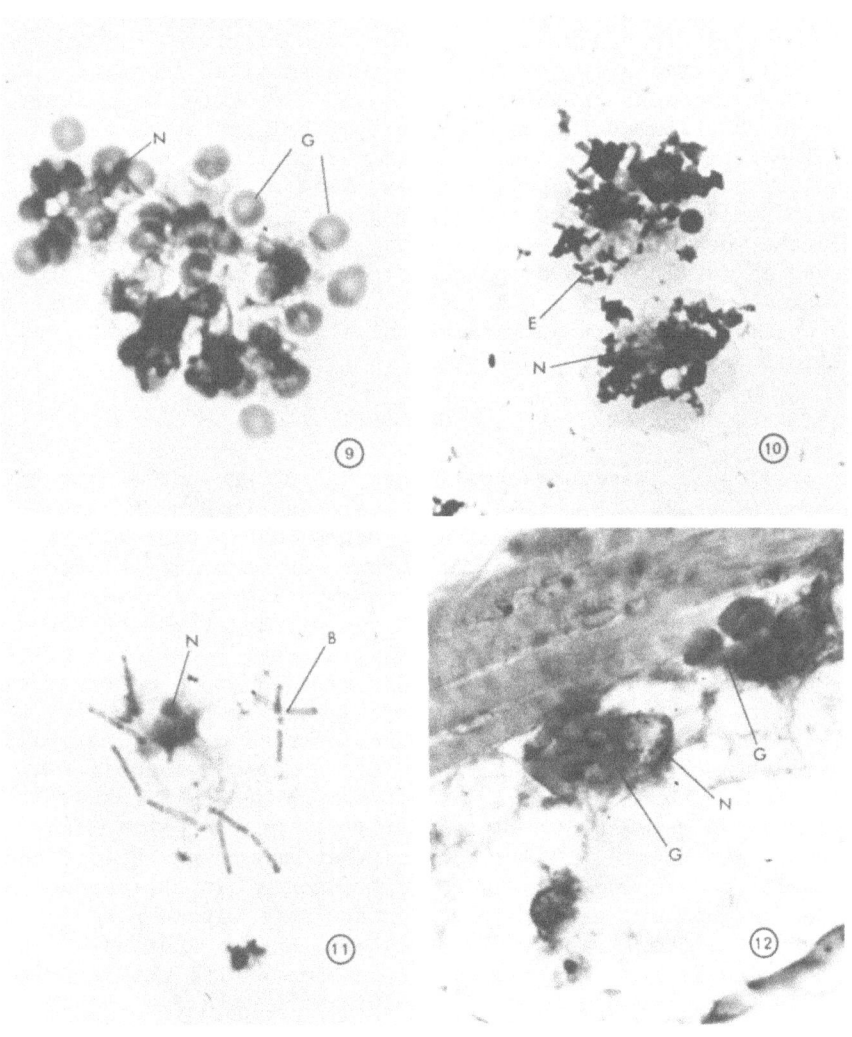

It was apparent from these experiments that the hemocytes of adult *A. stephensi* are capable of interacting with foreign particulate materials both *in vitro* and *in vivo*, although the demonstration of the interaction *in vitro* is much more striking than that of the interaction *in vivo*. This may be due to an actual lower responsiveness of the hemocytes in the *in vivo* environment, or to other as yet undetermined factors. Phagocytosis of foreign particulate materials has been reported to occur within the hemocoels of other mosquitoes. Hemocytes of larval *Anopheles quadrimaculatus* were capable of phagocytosing carmine particles from the hemolymph, but did not participate in tissue digestion during metamorphosis (Jones, 1954). Phagocytes in the mosquito hemocoel will apparently respond to bacterial infection within the mosquito. In a *C. pipiens* mosquito, Huff (1934) observed an infiltration of phagocytes into an area of the stomach wall which was undergoing a bacterial invasion. The same mosquito was infected with plasmodial oocysts, which, however, did not elicit any phagocyte response.

B. HEMOCYTE INTERACTION WITH SPOROGONIC STAGES OF PLASMODIUM

The previously described experiments had shown that *A. stephensi* possessed a moderate number of hemocytes, and that these hemocytes were capable of interaction *in vitro* and *in vivo* with foreign particulate materials. The next series of experiments was designed to determine if these hemocytes were capable of interaction *in vitro* or *in vivo* with sporogonic stages of the malaria parasite, in the same way as with other foreign materials.

The opinions of previous workers are divided as to whether or not mosquito hemocytes interact with the malaria parasite *in vivo*. Huff (1934) reported that no response of *C. pipiens* phagocytes to the sporogonic development of the malaria parasite *(Plasmodium cathemerium* and *P. relictum)* was observed by him during studies on mosquito susceptibility. On the other hand Weathersby and McCall (1968) observed that about one half of the ectopically developing oocysts *(P. gallinaceum)* in the hemocoels of *Aedes aegypti* or refractory *C. pipiens* were within phagocytic cells but that there was little or no evidence to indicate that these cells were inhibiting the development of the parasite. A "tissue rejection phenomenon" involving the malarial oocyst was described by Yoeli (1973) in *A. quadrimaculatus* infected with *Plasmodium berghei*. However, Yoeli did not describe any hemocytes which may have been participating. The characteristics of the rejection phenomenon in this unsuitable host included coagulation of the oocyst structure, abnormal thickening of the oocyst wall, development of hyaline demarcation zones around oocysts, and displacement of oocysts from surrounding tissues.

Although Strome and Beaudoin (1974) did not describe a hemocytic reaction to oocysts of *P. gallinaceum* and *P. berghei* on the
midgut walls of *A. aegypti* and *A. stephensi*, their scanning
electron micrographs of mosquito midguts infected with oocysts
included several spherical bodies on the midgut surface which
they labelled as hemocytes, including one *A. aegypti* hemocyte
on the surface of a *P. gallinaceum* oocyst. The significance of
this isolated observation was not discussed.

There have been no previous studies on the interaction *in
vitro* of mosquito hemocytes with the malaria parasite. Although
it is now possible to obtain sufficient mosquito hemocytes to
perform experiments of this nature, it is not yet possible to
obtain purified sporogonic stages of the malaria parasite.
Because of their intimate association *in vivo* with the mosquito
midgut or salivary glands, preparations of ookinetes, oocysts,
and sporozoites are nearly always contaminated with mosquito
tissue debris as well as with bacteria and yeast. Although
some purification of ookinetes (Weiss and Vanderberg, 1976) and
sporozoites (Mack *et al.*, 1978) is now possible, there are severe
technical limitations on the purity of these samples, and they
are of unsatisfactory quality for use in interaction experiments
in vitro.

Nevertheless, it was possible to complete some preliminary
studies on the interaction *in vitro* of malaria sporozoites and
ookinetes with mosquito hemocytes. It was difficult to quantitate these experiments because of the large amount of contaminating material in the parasite preparations, so I shall present
the results of these experiments in a qualitative manner only.

C. OOKINETES

Ookinetes of *P. berghei* were kindly prepared for these experiments by Mr. Martin Weiss using the method of Weiss and Vanderberg (1976). The ookinete suspensions were added to hemocoel
perfusates of mosquitoes on glass slides and permitted to stand
for 15 min at room temperature in a moist chamber. Afterwards,
these preparations were fixed with 2.5% glutaraldehyde in BSS
and stained with Giemsa's stain as described above.

Examination of these preparations revealed that hemocytes
interact *in vitro* with *P. berghei* ookinetes (Fig. 13).
Ookinetes were observed which had adhered to the surfaces of
several spread hemocytes. In addition, because the suspensions
of ookinetes included also free merozoites and trophozoites,
it was possible to observe the interaction of the mosquito
hemocytes with these stages as well. Many hemocytes were observed with free merozoites and trophozoites adhered to their
surfaces.

Figs. 13-14. Photomicrographs of hemocytes of *Anopheles stephensi* which have interacted *in vitro* with ookinetes or sporozoites of *Plasmodium berghei* (glutaraldehyde-fixed, Giemsa's stain). O = ookinete, S = sporozoite.

Fig. 13. Thinly spread hemocyte with ookinete.
Fig. 14. Thinly spread hemocyte with sporozoite.

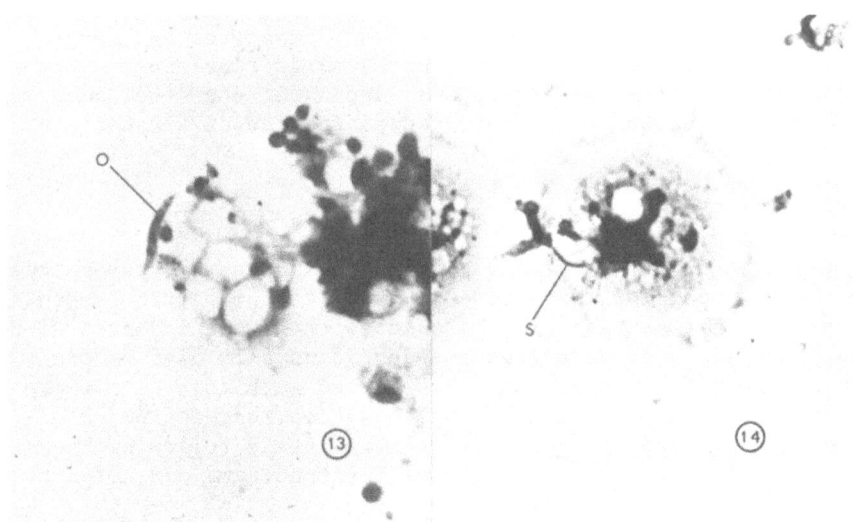

D. *SPOROZOITES*

Salivary gland sporozoites were obtained for these experiments from glands dissected from infected mosquitoes. The sporozoite suspensions were added to hemocoel perfusates and the preparations permitted to stand for 15 min at room temperature in a moist chamber. The preparations were subsequently fixed and stained as described above. Hemocytes in these preparations interacted with sporozoites. Several spread hemocytes were observed on which sporozoites were adherent (Fig. 14).

Although it was not technically feasible to quantitate these preliminary experiments, it was possible to observe a qualitative interaction *in vitro* between *A. stephensi* hemocytes and the sporogonic stages of *P. berghei*. However, further experiments *in vitro* with sporogonic stages will depend on the development of techniques for the satisfactory purification of these stages of the parasite.

E. *EXPERIMENTS in vivo*

Attempts to inject partially purified ookinetes or sporozoites into the hemocoels of mosquitoes to study the hemocytic response to them were not successful. Mosquitoes that were injected with these suspensions of parasites and debris usually died within the first hour post-injection. Associated with this high post-injection mortality was a rapid melanization reaction which occurred in the injected mosquitoes immediately following injection. Injected mosquitoes became moribund, and did not recover from anaesthesia.

CONCLUSIONS

It is possible at this point to summarize some of the findings presented here and to suggest a simple hypothesis which might explain the success of the sporozoite migration in the susceptible mosquito host. Briefly, *A. stephensi* possesses hemocytes capable of interacting *in vitro* with foreign particles as well as sporogonic stages of the malaria parasite. Although it was not possible to demonstrate hemocyte interaction *in vivo* with sporogonic stages, it was possible to demonstrate hemocyte interaction *in vivo* with GRBC. Why then is it possible for sporozoites to migrate through the hemocoels of these same mosquitoes to reach and penetrate the cells of the salivary glands?

Although hypotheses involving active evasion of the host defense mechanism may be attractive, it is not necessary to evolve a complicated scheme to explain the facts of the mosquito malarial infection. Quite simply, no matter what other factors may be operating, the mosquito cellular defense mechanism is numerically

Table 6. Interaction *in vitro* of *Anopheles stephensi* Hemocytes
          with Heat-killed Bacteria and Glutaraldehyde-fixed
          Red Blood Cells.[a]

Age of Mosquitoes (Days Post-Emergence)	Blood Meal[b]	Particle Suspension	% of Hemocytes Associated With At Least One Particle
0	None	*Escherichia coli*	>95%
0	None	*Bacillus subtilis*	>95%
0	None	Glutaralde-hyde-fixed RBC	>95%
14	Infected	Glutaralde-hyde-fixed RBC	>95%

a Time of interaction=15 min, temperature 21-23°C.

b At 3 days post-emergence, mosquitoes were permitted to feed
   upon a hamster infected with *Plasmodium berghei*.

overwhelmed by the parasite during a short time period and during the one phase in which the parasite is in direct contact with the extracellular hemolymph and hemocytes. The adult mosquito, with at most approximately 10,000 hemocytes (Table 5), is not equipped to deal with the prolific malaria parasite which may flood all the tissue spaces of the mosquito with a virtual sepsis (Oelerich, 1967) of sporozoites. Observations on the actual numbers of parasites that may be present vary, but it is useful to illustrate the argument with some examples. In *A. quadrimaculatus*, oocysts of *P. berghei* on the midgut average 60 and may reach a maximum of 400-500 (Yoeli, 1973). A single oocyst of *P. falciparum* may contain up to 9,555 sporozoites (Pringle, 1965). Thus, in a susceptible mosquito infected with a single oocyst, the number of sporozoites that may be released into the hemocoel may closely approximate the total number of hemocytes available in the entire mosquito. The success of the migration from the midgut to the salivary glands is evident from the numbers of sporozoites found in salivary glands of infected mosquitoes. In the salivary glands of a single mosquito, there may be 40,000 (Yoeli, 1973) or up to 200,000 (Shute and Maryon, 1966) sporozoites.

Further complications of the malarial infection could conceivably compromise the host even more greatly. If bacterial invasion of the stomach wall should occur as a result of disruption of its integrity by the developing oocyst, then the hemocytes that respond to this acute emergency (Huff, 1934) would be removed from the available pool of hemocytes, leaving less hemocytes to respond to the migrating sporozoites.

In the susceptible mosquito host, the overwhelming nature of the sporozoite migration might overshadow a possibly strong but futile reaction by the hemocytes. For the same reason, it might be difficult to observe active suppression of the cellular defense mechanism of the mosquito which was mediated by the malaria parasite. The *in vitro* methods described in this report offer an alternative means of studying interactions between hemocytes and parasite. However, the full utilization of these methods must await technological advances in the preparation and purification of sporogonic stages of the malaria parasite.

ACKNOWLEDGMENTS

I would like to thank Dr. Jerome Vanderberg for his willingness for me to use some of our collaborative data on the hematology of *Anopheles stephensi* which we have not yet published. In the same vein, I would like to thank Dr. Stephen Mack for the use of our as yet unpublished data on the hemolymph of mosquitoes. I would also like to thank Mr. Martin Weiss for making available to me his concentrated preparations of ookinetes. The original research in this report was supported in part by National Institutes of Health Research Service Award No. 5F32-A105221-02, as well as the U.S. Army Medical Research and Development Command under Contract No. DADA 17 73 C 3027. Additional support was from the U.S. National Institutes of Health through research Grant No. Al 09560.

REFERENCES

Amouriq, L. (1960). Formules hemocytaires de la larve, de la nymphe et de l'adulte de *Culex hortensis* (Dipt. Culicidae). *Bull. Soc. Entom. France*, 65, 135-139.

Andreadis, T. G. and Hall, D. W. (1976). *Neoaplectana carpocapsae*: Encapsulation in *Aedes aegypti* and changes in host hemocytes and hemolymph proteins. *Exp. Parasitol.*, 39, 252-261.

Arnold, J. W. (1974). The hemocytes of insects. *In:* "The Physiology of Insecta," (M. Rockstein, ed.), second edition, volume 5, pp. 201-254. Academic Press, New York.

Bhat, U. K. M. and Singh, K. R. P. (1975). The haemocytes of the mosquito *Aedes albopictus* and their comparison with larval cells cultured *in vitro*. *Experientia*, 31, 1331-1332.

Bronskill, J. F. (1962). Encapsulation of rhabditoid nematodes in mosquitoes. *Can. J. Zool.*, 40, 1269-1275.

Chao, J. and Ball, G. H. (1956). Quantitative microinjection of mosquitoes. *Science*, 123, 228-229.

Foley, D. A. and Cheng, T. C. (1977). Degranulation and other changes of molluscan granulocytes associated with phagocytosis. *J. Invertebr. Pathol.*, 29, 321-325.

Huff, C. G. (1934). Comparative studies on susceptible and insusceptible *Culex pipiens* in relation to infections with *Plasmodium cathemerium* and *P. relictum*. *Am. J. Hyg.*, 19, 123-147.

Jones, J. C. (1954). The heart and associated tissues of *Anopheles quadrimaculatus* Say (Diptera:Culicidae). *J. Morph.*, 94, 71-125.

Jones, J. C. (1962). Current concepts concerning insect hemocytes. *Am. Zool.*, 2, 209-246.

Jones, J. C. (1970). Hemocytopoiesis in insects. *In:* "Regulation of Hematopoiesis," (A. S. Gordon, ed.), volume I, pp. 7-65. Appleton-Century-Crofts, New York.

Mack, S., Foley, D. A., and Vanderberg, J. P. (1978). Hemolymph volume of *Anopheles stephensi*. (manuscript in preparation).

Mack, S., Vanderberg, J. P., and Nawrot, R. (1978). Column separation of *Plasmodium berghei* sporozoites. *J. Parasitol.* (in press).

Oelerich, S. (1967). Vergleichende Untersuchungen uber das Auftreten von Malaria-Sporozoiten in den Speicheldrusen und in den ubrigen Organen der Mucke. *Z. Tropenmed. Parasit.*, 18, 285-303.

Pringle, G. (1965). A count of the sporozoites in an oocyst of *Plasmodium falciparum*. *Trans. R. Soc. Trop. Med. Hyg.*, 59, 289-290.

Shute, P. G. and Maryon, M. E. (1966). "Laboratory Technique for the Study of Malaria," second edition, 112 pp. J. and A. Churchill, Ltd., London.

Strome, C. P. A. and Beaudoin, R. L. (1974). The surface of the malaria parasite 1. Scanning electron microscopy of the oocyst. *Expt. Parasitol.*, 36, 131-142.

Vanderberg, J. P. (1977). *Plasmodium berghei:* Quantitation of sporozoites injected by mosquitoes feeding on a rodent host. *Expt. Parasitol.*, 42, 169-181.

Weathersby, A. B. and McCall, J. W. (1968). The development of *Plasmodium gallinaceum* Brumpt in the hemocoels of refractory *Culex pipiens pipiens* Linn. and susceptible *Aedes aegypti* (Linn.). *J. Parasitol.*, 54, 1017-1022.

Weiss, M. M. and Vanderberg, J. P. (1976). Studies on *Plasmodium* ookinetes. 1. Isolation and concentration from mosquito midguts. *J. Protozool.*, 23, 547-551.

Whitcomb, R. F., Shapiro, M., and Granados, R. R. (1974). Insect defense mechanisms against microorganisms and parasitoids. *In:* "The Physiology of Insecta," (M. Rockstein, ed.), second edition, volume 5, pp. 447-536. Academic Press, New York.

Yoeli, M. (1973). *Plasmodium berghei:* Mechanisms and sites of resistance to sporogonic development in different mosquitoes. *Expt. Parasitol.*, 34, 448-458.

The Inducible Immunity System of Giant Silk Moths

H. G. BOMAN
I. FAYE
A. PYE[1]
T. RASMUSON

Department of Microbiology
University of Stockholm
S-106 91 Stockholm, Sweden

[1]Present address: Department of Zoophysiology, University of Umeå, S-901 87 Umeå 6, Sweden

I. INTRODUCTION TO THE SYSTEM

In this paper we intend to summarize briefly the main findings
from our work on the inducible defense system of giant silk
moths. This immunity manifests itself as a potent, cell-free
antibacterial activity in the hemolymph (Boman *et al.*, 1974a).
A key tool for us has been an *in vitro* assay of bacterial killing
measured by the loss of the colony forming ability. Our main
test organism has been *Escherichia coli*, D31 (Boman *et al.*, 1974).
This strain carries resistance to streptomycin and ampicillin,
properties which we have used both to eliminate any contaminating
bacteria and for selective plating during double infections.
Some of the parameters of our assay as well as its antibacterial
spectrum have recently been discussed elsewhere (Rasmuson and
Boman, 1977).

Depending on the supply of insects, we have worked with pupae
of *Samia cynthia*, *Callosamia promethea*, and *Hyalophora cecropia*.
However, the last species is the best model system, not only
because of its size and its stable diapause, but also because of
the amount of information already available. During diapause,
when the metabolism of the pupae is reduced to a few per cent
of the normal rate, it is still possible to induce killing
activity. Therefore, this condition offers an opportunity to
label the products made by genes controlling immune functions
without having much background from other biosynthetic reactions.
The use of labelled amino acids in combination with the bioassay
and methods for protein fractionation, has been our main approach.
In addition, *in vivo* and *in vitro* experiments with different
inhibitors have been of much help for our present understanding
of the system.

II. INDUCTION REQUIRES RNA AND PROTEIN SYNTHESIS

Induction of the antibacterial activity is accomplished by an
injection of viable, nonpathogenic bacteria like *E. coli*,
Bacillus subtilis, or *Enterobacter cloacae* (Boman *et al.*, 1974a;
Faye *et al.*, 1975). However, the latter organism is the most
effective vaccine yet found. Bacteria killed by freezing or
ultraviolet radiation, isolated cell walls, or preparations of
outer membrane or lipopolysaccharide (LPS) were always less
effective and gave somewhat irregular results. The response is
apparently unspecific; each bacterium tried will induce killing
activity against a variety of bacteria (cf. Rasmuson and Boman,
1977). We have recently investigated the fate of the immunizing
bacteria and found that they are phagocytized by cells in the
fat body predominantly located in both ends of the pupae (Faye,
1978). However, it remains to be determined how a nonspecific
process like phagocytosis is translated to a signal that specifically affects the genes of the immune system.

Studies of the induction kinetics demonstrated a lag period of some 8 hrs before the antibacterial activity appeared (Faye *et al.*, 1975). Maximum activity was usually recorded at days 2-4 and was followed by a slow decline. A second injection of bacteria gave no enhancing effect. We have also investigated the possibility of induction in pupae given inhibitors of RNA or protein synthesis. These results are summarized in Table 1. When pupae were given actinomycin D or cycloheximide, either together with the bacteria or soon after, the antibacterial activity was almost totally repressed (Faye *et al.*, 1975). These results indicated that *de novo* synthesis of RNA and protein were needed. If the inhibitors were given at peak height, there was no effect with actinomycin D while cycloheximide caused a rapid decline of the antibacterial activity. Thus, when RNA synthesis was completed, the messenger seemed to be stable, while the decline observed with cycloheximide indicated a turnover of one or more of the proteins involved.

The killing activity was found early to be sensitive to trypsin; since *de novo* synthesis of proteins was required, it was natural to examine the hemolymph proteins with SDS electrophoresis. Comparison of normal and immune hemolymph showed no clear differences, a fact which indicated that the immune proteins were only a minor part of the total blood proteins. We therefore had to make use of the selective labelling offered by the diapause and it was then found that the synthesis of 8 labelled polypeptides (P1-P8) was correlated with the appearance of the killing activity (Faye *et al.*, 1975).

Every injection of a pupa creates an injury which starts repair processes (Faye *et al.*, 1975; Telfer and Williams, 1960). Moreover, in nature an infection may often start with an injury. It was, therefore, necessary to investigate if the injury created by an injection of a physiological salt solution also induced any killing activity. This was found to be the case, but the activity was always significantly lower than that obtained with live bacteria (Fig. 1).

The injury effect and the fact that different bacteria gave the same killing activity prompted us to formulate the following two questions: (1) Does an infection and an injury cause the same response in protein synthesis? (2) Does an infection with a Gram-negative bacterium like *E. cloacae* and an infection with a Gram-positive bacterium like *B. subtilis* produce the same response in protein synthesis?

To answer these questions we treated three pupae of *H. cecropia* as follows: the first one was given an injection of ^{14}C-leucine and *E. cloacae*, the second one obtained ^{3}H-leucine and *B. subtilis*, while the third one got ^{3}H-leucine and a physiological salt solution (Faye *et al.*, 1975). At day 3 the pupae were sacrificed. The hemolymph from the first pupa was divided into two equal parts

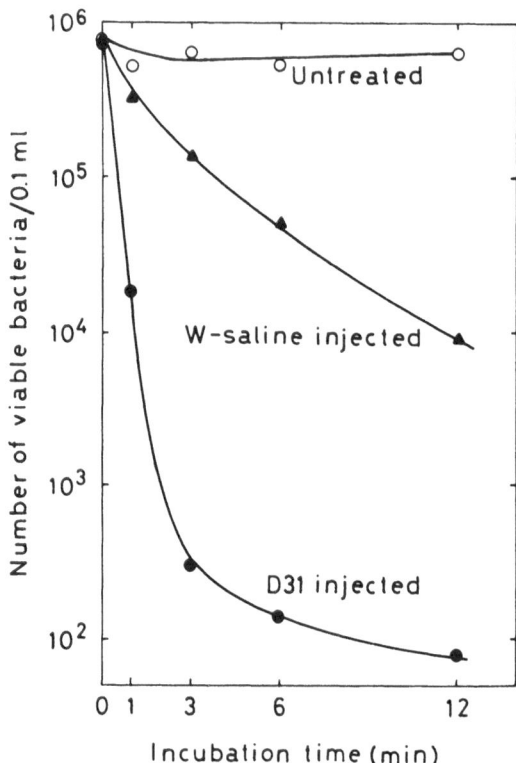

Fig. 1.   *In vitro* assay of the antibacterial activity in hemolymph
          from 3 different pupae of *Samia cynthia* pretreated as
          follows: pupa injected with 50 μl containing 5 x 10⁵
          viable cells of *E. coli*, D31 (●); pupa injected with
          50 μl of a sterile, physiological salt solution (W-saline)
          (▲); untreated pupa (o). The test organism was
          Escherichia coli, D31, the hemolymph concentration was 1%
          and incubation was at room temperature (Boman *et al.*,
          1974a.

Table 1. *In vivo* Effects of Inhibitors of RNA and Protein
Synthesis

Treatment of the pupae	Administration of inhibitor	
	Early	Late
Control	+	+
Actinomycin D (25 µg/pupa)	−	+
Cycloheximide (50 µg/pupa)	−	−

Six pupae of *Samia cynthia* were given actinomycin D or cycloheximide either 5 hr or 3 days after immunization (Faye *et al.*,
1975). The control obtained the same volume of a sterile salt
solution. Hemolymph samples were removed 48 hr after the
administration of the inhibitor and assayed for antibacterial
activity against *E. coli*, D31, High killing activity +, low or
none −.

and mixed with the same volume of blood from pupa number two and three, respectively. Electrophoretic studies of the isotope distribution in these mixed hemolymph samples showed that basically the same response was produced as the result of an infection with either *E. cloacae* or *B. subtilis* (Fig. 2). However, there was both a qualitative and a quantitative difference between the labelled polypeptides produced as the result of an infection with *E. cloacae* and the injury response created by an injection of a salt solution (Fig. 3). Thus, the pupae seem to have only an "on-or-off-switch" for infections with different bacteria. If the same "switch" also has a third position for the injury response cannot be determined at the present time.

III. FRACTIONATION AND RECONSTITUTION EXPERIMENTS

Early fractionation attempts using the killing assay indicated that this activity was due to a multicomponent system (Boman *et al.*, 1974a). Moreover, the proteins responsible for the antibacterial activity represent only a minor part of the total hemolymph proteins. Purification of the immune proteins would, therefore, have to depend on assays of both the antibacterial activity and the labelled amino acids incorporated. This dual assay approach was needed to establish if the labelled proteins (P1-P8) were directly responsible for the killing of bacteria. If so, reconstitution of the killing activity from purified labelled proteins should be possible. To approach this stage we had to start with fractionation of the hemolymph.

Our first separation attempts involved the use of cation and anion exchangers, but the material obtained from these chromatograms never gave any killing activity in reconstitution experiments. However, fractional precipitation with ammonium sulfate was found to give a workable system (Faye *et al.*, 1975). When three fractions (designated A, B, and D) were mixed, we could demonstrate that the killing activity was significantly higher than expected from assays of each of the fractions. We also discovered that the ratio of the components in reconstitution experiments is a crucial factor (Fig. 4). From this experiment we had to conclude that for an intelligent interpretation of the killing assay one must know if one and the same component is rate-limiting, and one must identify this factor. Since this still is not known, changes in a killing activity must be interpreted with caution (cf. Rasmuson and Boman, 1977).

Electrophoretic characterization of the labelled proteins in each of the ammonium sulfate fractions revealed that most of the radioactivity in the first precipitate was due to a single component, immune protein P5 (Faye *et al.*, 1975). We have made use of this fortunate circumstance and worked out an isolation procedure for P5 (Pye and Boman, 1977). The purified

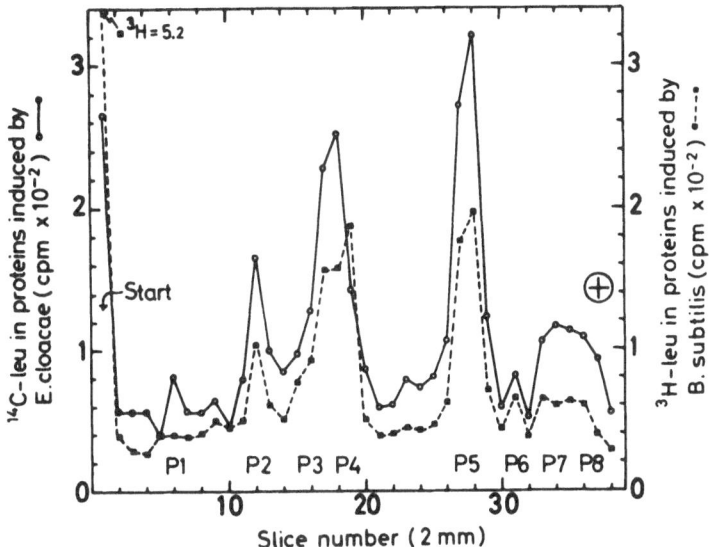

Fig. 2.   Co-electrophoresis of labelled polypeptides in hemolymph
          samples from 2 pupae of *Hyalophora cecropia*, one injected
          with ^{3}H-leucine and viable *Bacillus subtilis* (■---■), the
          other with ^{14}C-leucine and viable *Enterobacter cloacae*
          (o——o). The polyacrylamide gel concentration was 7.5%
          and the buffer contained 0.1% sodium dodecyl sulfate (SDS).
          Each slice of the gel was combusted in an oxidizer which
          separates the ^{3}H$_2$0 and the ^{14}CO$_2$ formed (Faye *et al.*,
          1975). The different proteins were tentatively designated
          P1-P8 as indicated from their respective mobilities. The
          purification of P5 was recently described (Pye and Boman,
          1977).

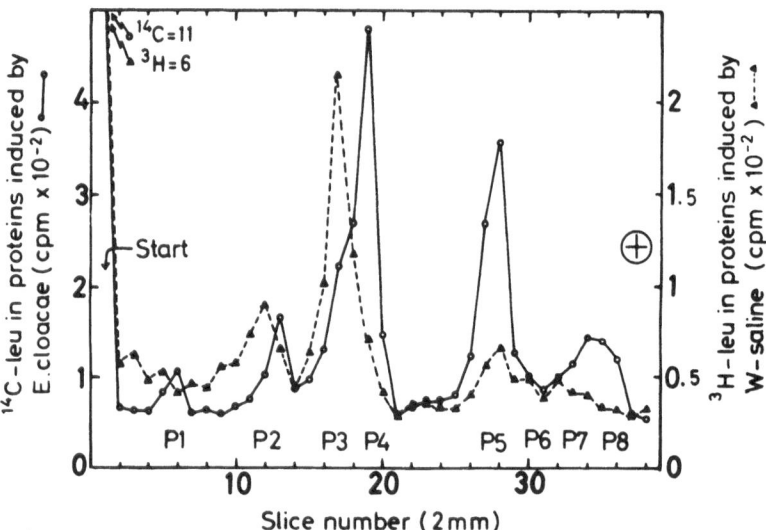

Fig. 3. Co-electrophoresis of labelled polypeptides in hemolymph
         samples from two *Hyalophora cecropia* pupae, one injected
         with ^{14}C-leucine and viable *Enterobacter cloacae* (o——o),
         the other injected with ^{3}H-leucine and sterile,
         physiological salt solution (▲---▲). Experimental
         details and processing were as described for figure 2.

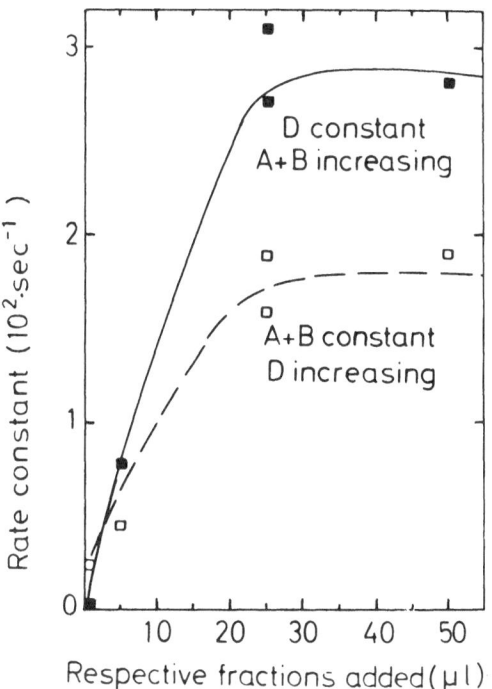

Fig. 4. Reconstitution of the antibacterial activity using three
hemolymph fractions (designated A, B, and D) obtained by
ammonium sulfate precipitations (Faye *et al.*, 1975). Each
point corresponds to a killing curve like the one showed
in figure 1. We have assumed a pseudo-first-order reaction
and calculated the apparent rate constants.

material was found to have a molecular weight of 96,000 and to be composed of four subunits of equal size. Protein P5 had no killing activity by itself, but it did enhance the killing activity of fractions B + D. However, the effect of P5 was only modest, and we expect fraction A to contain one or more additional factors which contribute to the killing activity. One such component ("the preexisting factor") has been identified in fraction Ac prepared from hemolymph of untreated pupae (Faye *et. al.*, 1975).

Protein P4 is the main labelled component in ammonium sulfate fraction D which, in addition, contains most of the lysozyme activity. Chromatography on sulfoethyl cellulose was used to separate protein P4 and the lysozyme, an experiment which also demonstrated that the latter component was essentially free from radioactivity (Fig. 5). The role of the lysozyme in defending insects against infections has been debated by several investigators (Mohrig and Messner, 1968; Kawarabata, 1971; Chadwick, 1975; Faye *et al.*, 1975; Powning and Davidson, 1976; Natori, 1977). Further purification and reconstitution experiments hopefully should settle the question.

IV. EVIDENCE FOR THREE SYSTEMS WITH DIFFERENT KILLING SPECIFICITIES

Some years ago we found that purified LPS was an effective inhibitor of the killing of *E. coli* while it did not affect the killing of *B. subtilis* (Fig. 6). The interpretation was obvious: the two bacteria must be killed by separate mechanisms. Later on we discovered that *Bacillus thuringiensis* produces two different immune inhibitors, tentatively designed A and B (Edlund *et al.*, 1976). Inhibitor A selectively blocks the killing of *E. coli* without affecting the activity against *B. cereus* while inhibitor B has the reverse effects. In addition, we have recently found that zymosan is an effective inhibitor which selectively blocks the killing of *B. cereus*. The known effects of these inhibitors are summarized in Table 2 which suggests that the *Cecropia* defense system is composed of at least three separate killing mechanisms, each one with a different specificity. However, it will be necessary to confirm this conclusion by further fractionations and reconstitutions of each of the systems.

Not much is known about the mechanisms of killing. With *E. coli* the reaction is quite fast and as far as we can tell killing follows single-hit kinetics (Boman *et al.*, 1974a). We also know that the osmotic pressure inside *E. coli* contribute to the rate of killing (Faye *et al.*, 1975). This effect could not be demon-

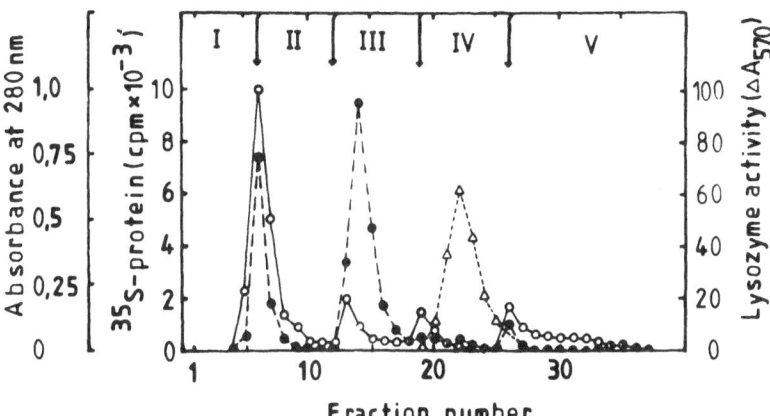

Fig. 5. Chromatographic separation of the labelled protein P4
          (•---•) and lysozyme (△---△) on a column of sulfoethyl
          cellulose. The starting material was an ammonium sulfate
          fraction (a 58-85% "cut") obtained from immune hemolymph
          of *Hyalophora cecropia*. The column was in equilibrium
          with 0.02 M potassium phosphate buffer which was also
          used as the first eluting agent, I. At tube 6 elution
          was shifted to 0.1 M potassium phosphate buffer con-
          taining 0.02 M KCl, II. At tube 12 elution was started
          with 0.1 M potassium phosphate buffer containing 0.1 M
          KCl, III. At tube 19 elution was begun with 0.5 M
          potassium phosphate buffer IV, and finally after tube
          26 elution was performed with 0.1 M KOH, V. Protein
          was monitored as A_{280} (o——o).

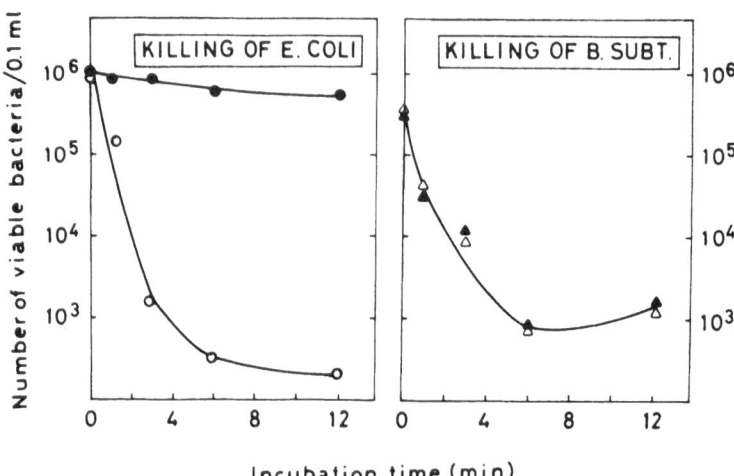

Fig. 6. Effect of LPS (filled symbols) on the *in vitro* killing
        of *Escherichia coli* D31 (circles) and *Bacillus subtilis*
        (triangles). Hemolymph was preincubated for 10 minutes
        with LPS from *Escherichia coli*, D31 at a concentration of
        500 μg/ml; the control (unfilled symbols) was preincu-
        bated with an equal volume of buffer. A sample from
        each incubation mixture was then added to standard
        reaction mixtures with the respective bacteria (Boman
        et al., 1974a). With *Escherichia coli*, D31 the final
        concentration of hemolymph was 1%, with *Bacillus
        subtilis* it was 5%.

Table 2.   Selective Effects of Four Microbial Inhibitors on
           the *in vitro* Killing of Three Bacteria.

Test Organism	Inhibitor			
	LPS	Zymosan	In. A	In. B
E. coli, D31	+	–	+	–
B. subtilis, Bsll	–	–	n.d	n.d
B. cereus, Bcll	–	+	–	+

The inhibitors were tested in our standard assay for antibacterial
activity (Boman *et al.*, 1974a).   The final conc. of LPS and
zymosan was 50 µg/ml.   Inhibitors A and B from *Bacillus thuringiensis* (Edlund *et al.*, 1976) are not yet purified.   Thus, their
respective concentrations could not be determined.   + indicates
more than 90% inhibition, – less than 10% inhibition in time
curves with at least 4 points.

strated for *B. subtilis* (Faye *et al.*, 1975). With *E. coli*, the
terminal step is lysis and this event follows a typical multi-
hit curve (Boman *et al.*, 1974a). Two lines of evidence indicate
that the lipid A moiety in LPS is the target in *E. coli*. The
tentative structure of LPS from *E. coli* D21 and 3 mutants is
given in Fig. 7 (data from Prehm *et al.*, 1976). Mild acid
hydrolysis of LPS is known to destroy the KDO part of the
molecule and produce free core polysaccharide and lipid A.
When these two parts were tried for inhibition of *in vitro* killing of *E. coli*, the polysaccharide part of the molecule showed
only low activity while the lipid A moiety was more active than
intact LPS (Faye *et al.*, 1975). The other line of evidence came
from a comparison of the killing of a series of interrelated
E. coli mutants which have lost increasing parts of their LPS
core (cf. Fig. 7). This experiment (Fig. 8) showed an increasing
susceptibility for strains which had lost increasing parts of
the core polysaccharide (Boman *et al.*, 1974a,b). These results
are consistent with the hypothesis that lipid A is the target
in *E. coli*. However, this claim should be substantiated by
direct binding studies which show that one or more of the immune
proteins bind to the lipid A moiety of LPS.

V. DISCUSSION OF OTHER SYSTEMS AND CURRENT LIMITATIONS

Humoral immunity mechanisms in insects were recently reviewed
by Chadwick (1975) who emphasized the need to correlate work
done with different systems. A necessary step on this road
would be the general use of the same test organisms and similar
assay conditions. We, therefore, emphasize that our test bacteria
are freely available to other investigators (*E. coli*, D31 as
well as the LPS mutants are also available from The *E. coli*
Genetic Stock Center, Yale University, New Haven, Connecticut
06510, USA).

Despite the lack of comparable assays we feel that our system
must be closely related to those described by Stephens and
Marshall (1962), Chadwick (1975) and Kawarabata (1970,1971).
It seems at present more difficult to relate our findings to
earlier studies by Briggs (1958), Gingrich (1964), Hink and
Briggs (1968) and Hink (1970).

We have not found any detectable antibacterial activity in the
hemolymph of untreated pupae or larvae of *H. cecropia* or *S.
cynthia* (see control in Fig. 1). This is contrary to the results
of Kinoshita and Inoue (1977) who recently reported that pooled
cell-free hemolymph from more than 2000 larvae of untreated
Bombyx mori contained antibacterial activity. However, to exclude species differences as well as possible infections of the
larvae it would be desirable to investigate if the activity
studied by Kinoshita and Inoue is different from the inducible
system one can expect to find in *B. mori*.

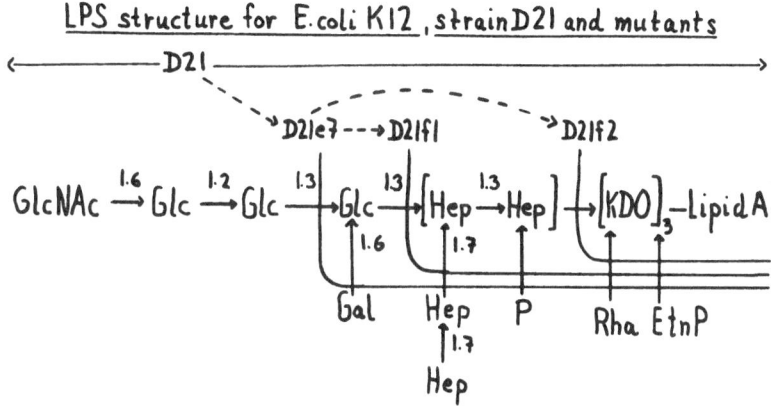

Fig. 7. Tentative structure for LPS from the parental strain
Escherichia coli, D21 and 3 spontaneous, consecutive
mutants D21e7, D21f1, and D21f2. The relationship
between the strains is indicated by the broken line
arrows in the upper part of the figure. A full account
of the structure work was published elsewhere (Prehm
et al., 1976). Brackets indicates that for Hep and KDO
it is not known to which unit the respective residues
are linked.

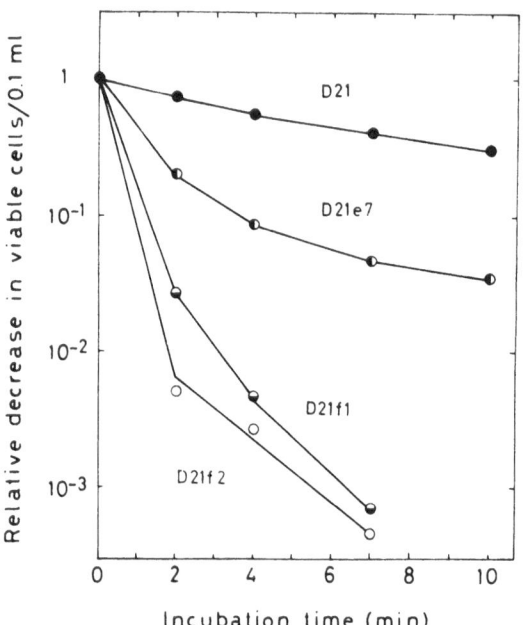

Fig. 8.   Susceptibility of four interrelated strains of
          Escherichia coli to killing by hemolymph from a pupa of
          Samia cynthia immunized with *Escherichia coli*, D31.
          The parental strain, D21 has LPS with a wild-type
          composition (●) while the mutants D21e7 (◐), D21f1
          (◑), and D21f2 (○) have lost increasing parts of their
          core polysaccharides as indicated in figure 7.

We would like to conclude by a short discussion of some factors which limit our present understanding of the *Cecropia* system. The most important bottleneck is quite clearly the killing assay. Elsewhere we have recently discussed some of the parameters which affect this bioassay (Rasmuson and Boman, 1977). Here, we would only like to emphasize that when we follow viability of a bacterium, we assay for the disappearance of a "substrate"; and we know nothing about the chemical products which are formed. Other short-comings stem from the fact that we have a multicomponent system without knowing the rate-limiting factor. We hope to improve the present situation by using gradually a cleaner system together with antisera prepared against the purified factors. However, the ultimate aim must be to describe in chemical terms how the different immune components interact with their targets and how they exert their lethal action on the different bacteria.

ACKNOWLEDGMENT

The work in Umeå as well as in Stockholm was supported by grants from The Swedish Natural Science Research Council (Dnr B2453). Recent acquisitions of equipments for the laboratory in Stockholm were made possible by a grant from The Knut and Alice Wallenberg Foundation.

REFERENCES

Boman, H. G., Nilsson-Faye, I., Paul, K., and Rasmuson, T, Jr. (1974a). Insect immunity. I. Characteristics of an inducible cell-free antibacterial reaction in hemolymph of *Samia cynthia* pupae. *Infect. Immun.*, 10, 136-145.
Boman, H. G., Nilsson-Faye, I., and Rasmuson, T, Jr. (1974b). Why is insect immunity interesting? p. 103-114. *In:* "Lipmann Symposium: Energy, Biosynthesis and Regulation in Molecular Biology," (D. Richter, ed.). Walter de Gruyter Verlag, Berlin.
Briggs, J. D. (1958). Humoral immunity in lepidopterous larvae. *J. Exp. Zool.*, 138, 155-188.
Chadwick, J. S. (1975). Hemolymph changes with infection or induced immunity in insects and ticks, p. 241-271. *In:* "Invertebrate Immunity," (K. Maramorosch and R. E. Shope, eds.) Academic Press, New York.
Edlund, T., Siden, I., and Boman, H. G. (1976). Evidence for two immune inhibitors from *Bacillus thuringiensis* interfering with the humoral defense system of Saturniid pupae. *Infect. Immun.*, 14, 934-941.
Faye, I., Pye, A., Rasmuson, T., Boman, H. G., and Boman, I. A. (1975). Insect immunity. II. Simultaneous induction of antibacterial activity and selective synthesis of some hemolymph

proteins in diapausing pupae of *Hyalophora cecropia* and
Samia cynthia. *Infect. Immun.*, 12, 1426-1438.

Faye, I. (1978). Insect immunity: Early fate of bacteria injected in *Saturniid* pupae. *J. Invertebr. Pathol.* 31, 19-26.

Gingrich, R. E. (1964). Acquired humoral immune response of
the large milkweed bug, *Oncopeltus fasciatus* (Dallas),
to injected materials. *J. Insect. Physiol.*, 10, 179-194.

Hink, W. F. (1970). Immunity in insects. *Transplantation
Proceedings*, 2, 233-235.

Hink, W. F. and Briggs, J. D. (1968). Bactericidal factors
in hemolymph from normal and immune wax moth larvae,
Galleria mellonella. *J. Insect. Physiol.*, 14, 1025-1034.

Kawarabata, T. (1970). Studies of an acquired resistance on
microbial infections in the silkworm (*Bombyx mori* L.). *J.
Facult. Agric.*, Kyushu Univ., 24, 231-254 (in Japanese with
English summary).

Kawarabata, T. (1971). Effect of vaccines on the production
of bactericidal activity and hemolymph lysozyme level in
the silkworm (*Bombyx mori* L.). *J. Facult. Agric.*, Kyushu
Univ., 16, 511-517.

Kinoshita, T. and Inoue, K. (1977). Bactericidal activity of
the normal, cell-free hemolymph of silkworms (*Bombyx mori*).
Infect. Immun., 16, 32-36.

Mohrig, W. and Messner, B. (1968). Immunreaktionen bei
Insekten. I. Lysozym als grundlegender antibakterieller
Faktor im humoralen Abwehrmechanismus der Insekten. *Biol.
Zbl.*, 87, 439-470.

Natori, S. (1977). Bactericidal substances induced in the hemolymph of *Sarcophaga peregrina* larvae. *J. Insect Physiol.*,
in press.

Powning, R. F. and Davidson, W. J. (1976). Studies on insect
bacteriolytic enzymes-II. Some physical and enzymatic properties of lysozyme from hemolymph of *Galleria mellonella*.
Comp. Biochem. Physiol., 55B, 221-228.

Prehm, P., Stirm, S., Jann, B., Jann, K., and Boman, H. G.
(1976). Cell-wall lipopolysaccharides of ampicillin-
resistant mutants of *Escherichia coli* K-12. *Eur. J. Biochem.*,
66, 369-377.

Pye, A. E. and Boman, H. G. (1977). Insect immunity. III.
Purification and partial characterization of immune protein
P5 from hemolymph of *Hyalophora cecropia* pupae. *Infect.
Immun.*, 16, 408-414.

Rasmuson, T. and Boman, H. G. (1977). The assay and the
specificity problem in insect community. *In:* "Developmental
Immunobiology," (J. B. Solomon and J. Horton, eds.). North-
Holland Biomedical Press, Amsterdam.

Stephens, J. S. and Marshall, J. H. (1962). Some properties of
an immune factor isolated from the blood of actively
immunized wax moth larvae. *Can. J. Microbiol.*, 8, 719-725.

Telfer, W. H. and Williams, C. M. (1960). The effects of diapause, development, and injury on the incorporation of radioactive glycine into the blood proteins of the *Cecropia*
silkworm. *J. Insect. Physiol.*, 5, 61-72.

INDEX